Mixed crop-livestock farming

A review of traditional technologies based on literature and field experiences

FAO
ANIMAL
PRODUCTION
AND HEALTH
PAPER

152

by
Hans Schiere
and
Loes Kater
La Ventana Agricultural Systems Analysis and Design
Bennekom, the Netherlands

Food
and
Agriculture
Organization
of
the
United
Nations

Rome, 2001

ISBN 92-5-104576-3

Foreword

Livestock play an important role in human society. In mixed farming systems in particular, they are able to utilize products that are not exploited by humans: kitchen wastes, grass from roadsides and wastelands, and crop residues from the cereal harvest. Animals give multiple products in return, such as meat, eggs, milk, fibres, social status and income, while dung and urine are valuable for fertilizing gardens, fields and fish ponds.

Traditional systems of animal keeping are not static. They adapt to changing circumstances such as increased population pressure, use of fertilizers and changing consumption patterns. Farmers, governments, and national and international agencies all have a role in this change, often by keeping track of new technologies and management practices that might be useful for development. The farmers' expertise may combine with the expertise of national and international agencies with access to information from different areas elsewhere in the world. The exchange of experiences between the grass-roots/farmer level and national and international level is necessary for the generation and application of appropriate technologies and management techniques that serve to keep farmers in business and to produce enough food for growing populations.

The Food and Agriculture Organization of the United Nations (FAO) has access to experiences regarding agricultural change across the world. Together with the Japanese Government it was decided to compile experiences from different locations, categorized by farming system, to make it easier for interested people to select ideas for their own conditions. This document presents a sample of such technologies for mixed farming systems, with emphasis on describing the livestock production systems in the traditional sector and identifying a sample of major constraints. For example, livestock production is constrained by poor management and inadequate research, extension and veterinary support services, as well as by ineffective communication between farmers and development agents. Suggestions for improving production in the livestock sector are given throughout the publication and a literature list is included for further reference.

A companion volume discusses traditional technologies for urban livestock systems. The combination of these two documents gives a broad overview of the technologies available for application under a variety of conditions.

Comments and suggestions for improvement from readers are welcomed to enable future upgrading of the material.

Contents

Tables

Figures

Acknowledgements

Thanks are due to the many people who helped to make this publication possible, in particular by providing suggestions and unpublished manuscripts. Particular mention is made of Roel Bosma (privatization of veterinary services), Juan Carlos Chirgwin, Rijk de Jong (rearing of young stock), Harm de Vries (rural poultry), Jaap de Winter (biogas), Peter Fellows (food products), Keith Hammond (breeding), Hannie Koch (veterinary aspects) and Cheryl Lans (indigenous veterinary medicine). The photographs have been provided by the authors, unless otherwise acknowledged.

This report is a product of the FAO-Japan Cooperative Project "Collection of Information on Animal Production and Health", with the support of the Japan Racing Association (JRA) fund.

Chapter 1

Introduction

Many farmers in tropical and temperate countries survive by managing a mix of different crops and/or animals. The best known form of mixed farming is when crop residues are used to feed the animals and the excreta from the animals are used as nutrients for the crops. Other forms of mixing take place where grazing under fruit-trees keeps the grass short, or where manure from pigs is used to "feed" the fish pond. Traditionally, a wide variety of mixed farming systems has been used worldwide. These systems are essential for the livelihood of farmers and for the production of food and other commodities for the cities and export markets. Even many highly specialized crop and livestock systems in developed and developing countries are rediscovering the advantages of mixed farming. For example, specialized industrial pig and poultry farmers are banned from modern countries such as Singapore, and in western Europe they are forced to exchange their dung surpluses with crop farmers. Moreover, the essence of many modern organic farming systems lies in the mixing of crops and animals.

What is the essence of mixed farming, how important is it, is it something of the past or is it a new vision for the future? Where and how does mixed farming occur and what can be done to make it more productive while at the same time giving due attention to aspects of sustainability? These are the main issues discussed in Chapters 2 and 3, which explain several forms of mixing and ways to characterize them on the basis of the relative availability of production factors – land, labour and capital. The similarity between issues of mixing in agriculture and in society is also apparent. In the same way that dung from animals is recycled for use on crops, it becomes increasingly important in cities to recycle newspapers, glass, tin cans, etc. Recycling can be necessary because of a lack of resources, such as in low external input agriculture (LEIA), and as a result of problems with waste disposal in high external input agriculture (HEIA).

Chapters 4 to 7 present a series of traditional technologies and their suitability under different conditions, together with examples of social organization to make various technologies successful. The existence of many different mixed farming systems and modes of production has at least two implications. The first refers to a wide variation in the types of technologies and management practices. The second refers to the fact that no one technology or a few technologies can be suitable for a wide range of systems. However, much can still be done in terms of traditional techniques for mixed farming, and these chapters focus on areas such as:

- selection of animal species (dairy animal, poultry);
- traditional disciplinary approaches (health, feeding, breeding, reproduction, etc.);
- typical interdisciplinary issues (dung and urine management, crop rotation, draught and soil conservation, use of biogas, agroforestry).

PHOTO 1
Conservation farming in Kenya: anti-erosion bunds are planted with grass that can be used to feed animals

PHOTO 2
A specialized dairy farm with buffaloes in India: these dung surpluses will be hard to recycle if the neighbourhood turns from rice fields into suburbs

The wide range of technologies for mixed farming has made it necessary to be selective, i.e. only some animal species are discussed in Chapter 4; only some aspects of disease, feeding and breeding are discussed in Chapter 5; nutrient cycling is discussed in Chapter 6 in a condensed way; while some aspects of social organization and management are given in Chapter 7. Chapter 8 presents a series of mixed farming systems that have proved to be successful over a longer period of time, together with some general conclusions.

The thinking behind this publication is based on modern insight from system theory and it is stressed that there is no such thing as one solution for all problems. Technologies and management methods can be very useful, but mostly only under certain conditions and circumstances. On the other hand, innovations to cope with changing farming systems are found everywhere. Not only can research provide new technologies, but the farmers themselves can also search for ways to cope with change. They modify traditional technologies and adopt and adapt methods passed on from colleagues and formal research centres.

Throughout it is aimed to show that the thinking and technicalities behind the mixing of several enterprises are things for the future, not only for the past, and not only for agriculture but also for society in general. Perhaps the biggest change that could or should take place to "improve" mixed systems is that farmers and development workers start to see the development of a part of the system in the context of the whole. This can be called the "communal ideotype", which implies that farmers and researchers should aim for maximum but sustainable output of the farm, sometimes at the expense of the output of an individual crop or animal. Another change is perhaps that official research recognizes farmers' experience and participation as assets to modify new technologies for their own variable conditions. Today this is already accounted for in many research and development projects that are implemented in developing countries but much more can be done.

Chapter 2
Characterization of mixed farms

Mixed farming is common worldwide, in spite of a tendency in agribusiness, research and teaching towards specialized forms of farming. Obviously, mixing has both advantages and disadvantages. For example, farmers in mixed systems have to divide their attention and resources over several activities, thus leading to reduced economies of scale. Advantages include the possibility of reducing risk, spreading labour and re-utilizing resources. The importance of these advantages and disadvantages differs according to the sociocultural preferences of the farmers and to the biophysical conditions as determined by rainfall, radiation, soil type and disease pressure. This chapter first describes several forms of mixing. Second, it explains how mixing of several parts requires a special approach to make a success of the total mix. What counts is the yield of the total, not of the parts. Trees in and on the edge of a crop field generally reduce the grain yield, but the combination of the trees (for fodder and timber) and crops is valuable, because each of the components produces useful products for the farm (people and animals included).

WHAT IS MIXED FARMING?

Mixed farming exists in many forms depending on external and internal factors. External factors are weather patterns, market prices, political stability, technological developments, etc. Internal factors relate to local soil characteristics, composition of the family and farmers' ingenuity. Farmers can decide to opt for mixed enterprises when they want to save resources by interchanging them on the farm – because these permit wider crop rotations and thus reduce dependence on chemicals, because they consider mixed systems closer to nature, or because they allow diversification for better risk management.

There is wide variation in mixed systems. Even pastoralists practise a form of mixed farming since their livelihood depends on the management of different feed resources and animal species. At a higher level, a region can consist of individual specialized farms and service systems that together act as a mixed system. Other forms of mixed farming include cultivation of different crops on the same field, such as millet and cowpea or millet and sorghum, or several varieties of the same crop with

---

BOX 1
IS MIXING AN IMPROVEMENT?

The choice of mixed farming is not always a sign of improvement of the situation in which people may find themselves. Mobile Fulani herdsmen in West Africa engage in crop production only when forced by circumstances, such as drought or animal diseases, leading to severe losses in livestock, making continuation of their former way of life impossible. Mixed farming is for them a poverty-induced option. Resource-poor farmers going into mixed farming have to apply labour-intensive techniques (their only resource) and, because of their low purchasing power, they cannot afford external inputs and have no option but to overexploit the environment.

(Based on Slingerland, 2000.)

---

different life cycles, which uses space more efficiently and spreads risks more uniformly (Photos 3 and 4).

The study of a wide variety of mixed systems at different levels is beneficial to understanding the logic of mixed systems in general. Disciplines such as ecology, economics and complex system theory have tools and concepts that can help us to understand better the mixed blessings of mixed systems. One essential point here is that the principle of mixing occurs everywhere, also in society – domestic waste such as glass, bottles or paper is also recycled. Another point is that in mixing the different functions of plants and animals can be observed: a cereal crop produces grain and straw, a legume provides grain, organic matter, fodder and nitrogen. A third point is that it tends to be more important to look for high yield of the combination of the components rather than for the (high) yield of one component. Mixed farms are systems that consist of different parts, which together should act as a whole. They thus need to be studied in their entirety and not as separate parts in order to understand the system and the factors that drive farmers and influence their decisions. That principle is here referred to as the "command ideotype"(Donald, 1981; Schiere *et al.*, 1999). It may be the most important principle to achieve increased production in mixed systems, together with the awareness that crops and animals have multiple functions (Boxes 2 and 3).

PHOTO 3
Mixed cropping: pyrethrum and maize (Kenya)

PHOTO 4
Mixed cropping: maize stover intersown with a legume (*Dolichos lablab*) to provide fodder after the grain harvest (Honduras)

FORMS OF MIXED FARMING

Mixed farming systems can be classified in many ways – based on land size, type of crops and animals, geographical distribution, market orientation, etc. Three major categories, in four different modes of farming, are distinguished here. The categories are:
- On-farm versus between-farm mixing (Box 4)
- Mixing within crops and/or animal systems (Box 5)
- Diversified versus integrated systems

The modes of farming refer to different degrees of availability of land, labour and inputs, ranging from plenty of land to a shortage of land. The modes are characterized by Schiere and De Wit (1995) as expansion agriculture (EXPAGR, plenty of land), LEIA, HEIA and new conservation agriculture (NCA, a form of land use where shortages are overcome by more labour, more inputs and keen management).

On-farm versus between-farm mixing

On-farm mixing refers to mixing on the same farm, and *between-farm mixing* refers to exchanging resources between different farms. On-farm mixing occurs par-

ticularly in LEIA where individual farmers will be keen to recycle the resources they have on their own farm. Between-farm mixing occurs increasingly in HEIA systems – in countries such as the Netherlands it is used to mitigate the waste disposal problems of specialized farming. Crop farmers use dung from animal farms, a process that involves transport and negotiation between farmers and even politicians. Between-farm mixing also occurs at the regional level – in the store cattle systems of the United Kingdom and the United States, animals are raised in one area to be fattened in another area where plenty of grain is available. In tropical countries also, manure may be transported from livestock farms to farmers and vegetable cropping areas where manure is in short supply.

Pastoralists from such systems in West Africa and on the Indian subcontinent also exchange cattle and crop products with crop farmers. Cultivators receive manure, labour and, less important, milk in return for cash, grain and water rights traded to pastoralists. Entrustment of livestock from crop farmers to pastoralists follows more or less the same rules. In return for taking care of the herd, herders receive either cash, or cropland, or labour for the cropland or a share of the milk and the offspring.

Mixing between nearby farms is considered here as providing the same advantages as on-farm mixing, but it should be underlined that there are important differences in terms of social organization and transaction costs. For example, in West Africa the exchange between farms leads to tension and accidents as crop farmers start to use land that used to be pastoral only. The amount of grazing land is decreasing and dependence on the grazing of crop residues is increasing. When herders bring in the animals before the field is properly harvested serious incidents and conflicts can arise.

BOX 4

BUFFALOES IN A SMALLHOLDER DAIRY IN THE HINDU KUSH-HIMALAYAS: A CASE OF BETWEEN-FARM MIXING

A farmer buys a milch buffalo from a lowland buffalo trader at a price of between Rs 23 000 and 28 000. The animal is milked for eight months to a year and the gross income amounts to Rs 25 000-30 000. Some farmers sell expectant *bakerno* buffaloes for Rs 18 000-20 000 after one year of milking and buy an in-milk buffalo *laino* from the trader at a price of Rs 25 000-30 000 to ensure continued milk production. If the farmer waits for the *bakerno* buffalo, it takes about eight months before it calves and begins producing milk. The strategy is most suitable for smallholders who manage only one buffalo, but farmers who have more than one buffalo also adopt it. The lactating buffaloes come from the lowland areas where conception is easier.

(Based on Tulachan and Neupane, 1999.)

Mixing within crop and/or animal systems

Mixing within crop and/or within animal systems refers to conditions where multiple cropping is practised, often over time, or where different types of animals are kept together, mostly on-farm. Both these systems occur frequently though they are not always apparent.

Within-crop mixing takes place where crop rotations are practised over and within years. For example, a farmer has a grain-legume rotation to provide the grain with nitrogen or a potato-beet-grain rotation to avoid disease in the potatoes. Plants can also be intercropped to take maximum advantage of light and moisture, to suppress weeds or prevent leaching of nutrients through the use of catch crops. Examples of mixing between animals are found in chicken-fish pond systems where chicken dung fertilizes the fish pond; in beef-pork systems where pigs eat the undigested grains from the beef cattle dung; or in mixed grazing such as cow-sheep mixes to maximize biomass utilization or to suppress disease occurrence (Photos 5 and 6).

PHOTO 5
Cows and sheep grazing together in a pasture in the Netherlands to optimize biomass utilization and to reduce disease pressure

PHOTO 6
A model farm near Durban (South Africa) where the dung from the chickens is used to fertilize the algae growth in the fish pond

BOX 5
MORE MIXING OF LIVESTOCK

By keeping several species, farmers can exploit a wider range of feed resources than if only one species is kept. In pastoral areas, camels can graze up to 50 km away from watering points, whereas cattle are limited to a grazing orbit of 10-15 km. Camels and goats tend to browse more, i.e. to eat the leaves of shrubs and trees; sheep and cattle generally prefer grasses and herbs. Different animal species supply different products; e.g. camels and cattle can provide milk, transport and draught power, whereas goats and sheep tend to be slaughtered more often for meat. Chickens often provide the small change for the household, sheep and goats are sold to cover medium expenditures, while larger cattle are sold to meet major expenditures.

Keeping more than one species of livestock is also a risk-minimizing strategy. An outbreak of disease may affect only one of the species, e.g. the cow, and some species or breeds are better able to survive droughts and thus help carry a family over such difficult periods. Advantage can also be taken of the different reproductive rates of different species to rebuild livestock holdings after a drought. For example, the greater fecundity of sheep and goats permits their numbers to multiply quicker than cattle or camels. The small ruminants can then be exchanged or sold to obtain large ruminants.

(Reijntjes, Haverkort and Waters-Bayer, 1992.)

Diversified versus integrated systems

The distinction between diversified and integrated systems is perhaps the most relevant one for this report. Diversified systems consist of components such as crops and livestock that co-exist independently from each other. In particular, HEIA farmers can have pigs, dairy and crops as quite independent units. In this case the mixing of crops and livestock primarily serves to minimize risk and not to recycle resources.

Integration is done to recycle resources efficiently. It occurs in mixed ecological farms of temperate countries (here called the mode of new conservation agriculture, NCA), but also in mixed, relatively low input farms of southern and southwestern Australia with grain-legume-sheep mixtures. Integration occurs most often, however, in LEIA farming systems that exist in many tropical countries where products or by-products of one component serve as a resource for the other – dung goes to the crops and straw to the animals. In this case the integration serves to make maximum use of the resources. Unfortunately, these systems tend to become more vulnerable to disturbance because mixing of resource flows makes the system internally more complex and interdependent.

In Asia, the integration of livestock, fish and crops has proved to be a sustainable system through centuries of experience. In China, for example, the integration of fishpond production with ducks, geese, chickens, sheep, cattle or pigs increased fish production by 2 to 3.9 times (Chen, 1996), while there were added ecological and economic benefits of fish utilizing animal wastes. Environmentally sound integration is ensured where livestock droppings and feed waste can be poured directly into the pond to constitute feed for fish and zooplankton. Livestock manure can be used to fertilize grass or other plant growth that can also constitute feed for fish. Vegetables can be irrigated from the fishponds, and their residues and by-products can be used for feeding livestock.

Grazing of livestock under plantation trees such as rubber, oil palm or coconut is a form of crop-livestock integration that is often found in Southeast Asia. Experiments in Malaysia with cattle and goats under oil palm showed better oil palm bunch harvest and comparable results were found where goats fed under rubber trees. In rubber and oil palm plantations in Malaysia, the integration of livestock to utilize the vegetative ground cover under the tree canopy increased overall production and saved up to 40 percent of the cost of weed control. Similarly, sheep helped to control weeds in sugar cane fields in Colombia. This suppressed the costs of herbicides, reduced the cost of weed control by half and provided additional income from meat production (FAO, 1995a). This also occurs where cows graze under coconuts (Photo 7).

BOX 6
INTEGRATED FARMING SYSTEMS

An experimental farm in Thailand maintains pigs and chickens, as well as a vegetable garden and a fish pond. Animal wastes are used for fertilizer, fish feed and biogas generation. Crop and human wastes are also added to the biogas unit. Liquid effluent from the biogas generator is used in the fishpond and solid residues on the garden. Periodically the locations of the garden and the pond are reversed, so residues from one serve as nutrients for the other. Little is wasted in such a system.

(Based on BOSTID, 1981.)

PHOTO 7
Crop and livestock integration: cattle grazing under coconut trees (Sri Lanka)

PHOTO 8
Crop-livestock integration: sheep grazing under tall-stemmed fruit trees (the Netherlands)

The best known type of integrated mixed farming is probably the case of mixed crop-livestock systems. Cropping in this case provides animals with fodder from grass and nitrogen-binding legumes, leys (improved fallow with sown legumes, grasses or trees), weeds and crop residues. Animals graze under trees or on stubble, they provide draught and manure for crops, while they also serve as a savings account (Figure 1). This kind of system using crops and ruminants such as cattle, buffaloes, sheep and goats is the focus of this publication. But even here it is necessary to further distinguish different systems (called "modes") as explained in the following section.

MIXED CROP-LIVESTOCK SYSTEMS, DIFFERENT MODES

Even in integrated systems the exchange of resources such as dung, draught and crop residues takes place in degrees that differ among the so-called modes of farming (Schiere and De Wit, 1995), based on the availability of land, labour and capital respectively (Table 1):
- Expansion agriculture (EXPAGR)
- Low external input agriculture (LEIA)
- High external input agriculture (HEIA)
- New conservation agriculture (NCA)

FIGURE 1
An outline of different resource flows in mixed crop-livestock systems

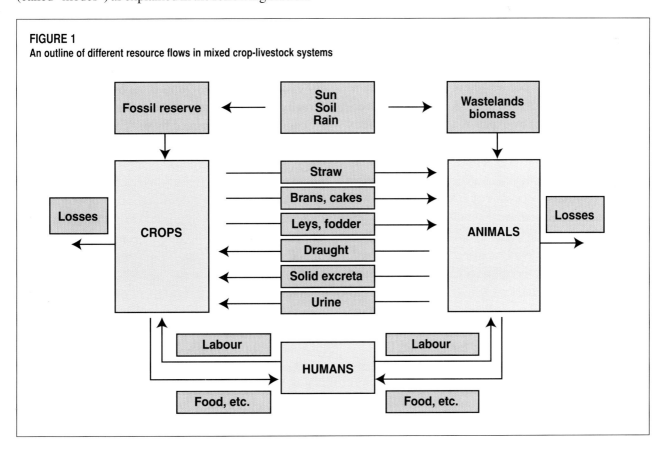

TABLE 1

Characterization of different modes of mixed crop-livestock farming

Mode of farming	EXPAGR	LEIA	HEIA	NCA
Relative access to production factors[1]:				
Land	+	-	-	-
Labour	-	+	-	+/-
Capital	-	-	+	+/-
Characteristics of farming:				
Source of animal feed	Outfield	Infield[2] roadsides	Infield Import	Infield
Role of animals as savings account	High	Medium	Low	Low
Importance of excreta - Dung - Urine	Positive Neglected	Positive Positive	Negative Negative	Positive Positive
Source of energy for labour	Humans/animals	Humans/animals	Fossil fuel	Fossil fuel/animals
Form of mixing	Diversity Can be between and on-farm	Integration On-farm	Specialization May be between farms	Integration Mainly on-farm
Crop residue feeding	Irrelevant	Very relevant	Irrelevant	Relevant
Role of leys				
- For weed control - For nutrient dynamics - For erosion control	NA[3] NA NA	Low/NA Low/NA Low/NA	NA NA Low/NA	Important Important Important
Ratio outfield/infield[2]				
- Local level - International level	High Low/NA	Low Low/NA	Low High	Low Low/NA
Output of milk or meat per animal	Low	Low	High	Medium
Attention to conservation of the resource base	Low	Medium	Low	High

[1] The access to land, labour and capital is to be read within a column, contrary to what has to be done for the comparison of system characteristics between modes (over rows). For example, a "-" for labour in the HEIA column means that labour is relatively scarce compared to capital inputs in that mode; not necessarily as compared with LEIA where it is indicated with a "+".
[2] Infield is defined as the crop area that depends on grazing from outfield for its nutrients.
[3] NA: not applicable.
Source: Based on Schiere and De Wit (1995).

Different modes of mixed farming

The EXPAGR mode occurs where land is abundant, i.e. where shortage of land or local fertility are overcome by migration or by expansion into other regions where bush and forest fallow still occur. Typical examples of mixed farming in this mode are found in West Africa and in old Asian and European grazing systems. Animals were sent out to graze and would (occasionally) come home to "pull the plough or fertilize the crop fields". The crop fields themselves could move elsewhere if local soil fertility declined. However, this mode is becoming more rare as land resources are exhausted throughout the world.

Mixed farming in LEIA occurs where the shortage of land can no longer be overcome by migration or use of substantial areas elsewhere for grazing. Lack of access to external inputs such as fuel, chemical fertilizers or pesticides implies that only increased use of labour and skills offers a way out. This also implies the introduction of modified practices, and the need to adjust demand according to resource availability (see, for example, Box 7). Dung is carried around on the farm by using more labour because a lack of soil fertility cannot be compensated by shifting to more land or by employing more livestock to "produce" more dung. In LEIA systems the latter is considered a resource but a waste product in HEIA systems (Photos 2, 9, 10 and 11). If not managed properly and if demand for food and other crops is not adjusted to the carrying capacity of the soil, this can result in mining of soils and/or collapse of the systems. Van der Pol (1992) calculated that the cotton-cereal systems in southern Mali earned 40 percent of their income by mining the soil. However, this cannot go on indefinitely and sooner or later the system will collapse if there are no changes. Some researchers think that animals, when managed correctly, can serve to fill part of the gap that exists between the output and the input of nutrients in the system, together with a proper use of chemical fertilizer.

PHOTO 9
Kenyan farmer holding dung in his hands; dung has become a valuable resource

PHOTO 10
A dung heap at an ecological farm in the Netherlands

PHOTO 11
The spout of a tank for separate urine collection from animals (the Netherlands)

PHOTO 12
Sheepsty on wasteland (the *Ginkelse hei*) in the Netherlands used to house animals and collect dung

Mixed farming in the HEIA mode is not frequently found because it implies plentiful access to resources such as external feed and fertilizer that make exchange and recycling of resources at farm level not relevant. Exchange of resources between farms only exists, as seen in the section *On-farm versus between-farm mixing*, after the excessive use of fertilizer forces farmers to recycle the waste. In the HEIA mode the demand for output determines the use of inputs. The use of external resources can reach such high levels that the environment is affected by emissions from the crop and/or animal production systems, ultimately leading to waste disposal problems, thus forcing HEIA into NCA.

NCA is a mode of farming where production goals are matched as closely as possible to the resource base. This approach represents a mix between HEIA and LEIA, i.e. it aims to replace the removed nutrients but it also aims to achieve keen farming and adjusted crop-

ping and consumption patterns to suit local conditions. The use of leys (improved fallows for grazing) is important to regenerate soils, to add nitrogen, to mobilize phosphate and to suppress weeds (i.e. to avoid herbicides).

CONCLUDING COMMENTS

Mixed systems occur in several forms. For example, pastoral systems have experience in the management of mixed herds and of livestock with feed resources. One form of mixing occurs where livestock is kept on grazing lands distant from cropland in the EXPAGR mode where land is abundant. Mixed systems can also occur as a combination of specialized farms that exchange resources among them, particularly in HEIA. This report focuses on the kind of mixing that is found in integrated crop-livestock systems. Diversified systems are a combination of specialized subsystems that

---

BOX 7
VILLAGE AGROFORESTS IN JAVA

Village agroforests have existed in Java since at least the tenth century and today comprise 15-50 percent of the total cultivated village land. They represent a permanent type of land use, which provides a wide range of products with a high food value (e.g. fruit, vegetables, meat and eggs) and other products, such as firewood, timber and medicines. In their small plots, often less than 0.1 ha, Javanese peasants mix a large number of different plant species. Within one village, up to 250 different species of diverse biological types may be grown: annual herbs, perennial herbaceous plants, climbing vines, creeping plants, shrubs and trees ranging from 10 to 35 m in height. Livestock form an important component of this agroforestry system – particularly poultry, but also sheep grazing freely or fenced in sheds and fed with forage gathered from the vegetation. The animals have an important role in nutrient recycling. Fish ponds are also common and the fish are fed with animal and human wastes. Natural processes of cycling water and organic matter are maintained; dead leaves and twigs are left to decompose, keeping a continual litter layer and humus through which nutrients are recycled. Compost, fishpond mud and green manures are used on cropland. These forms of recycling are sufficient to maintain soil fertility without the use of chemical fertilizers. Villagers regulate or modify the functioning and dynamics of each plant and animal within the system.

(Based on Reijntjes, Haverkort and Waters-Bayer, 1992.)

---

aim to reduce risk in conditions of variable but relatively abundant resources. Strong integration is associated with LEIA and NCA conditions where use of resources such as fertilizer and fossil fuel is restricted because of problems with pollution. This gives clues to development workers and policy-makers: cheap resources lead to specialization, restricted use of resources leads to mixing. An important aspect in promoting mixed farming is that the yield of the total enterprise is more important than the yield and/or efficiency of the parts. This is elaborated in the next chapter in which the technologies are presented.

Chapter 3

Traditional technologies – types and suitability

The previous chapter distinguished between different kinds of mixed farming. Such a classification is made to emphasize the variety in forms and functions of mixing. It also underlines that what is useful in one place may be counterproductive elsewhere. For example:

- Recycling of nutrients may be useful in LEIA and NCA systems but it is not likely to serve any practical purpose for farmers in EXPAGR and HEIA.
- Planting of fodders is important in HEIA but it is not likely to be of any use in EXPAGR because of the availability of feed on natural grazing lands. It may be difficult or impossible in LEIA due to a general shortage of land, but it is relevant in HEIA to provide good quality fodder. Fodder cultivation is useful in NCA to conserve soil, to be part of the crop rotation and to feed the animals.
- Housing and/or fencing is not practically useful in EXPAGR but it is part and parcel of livestock systems in LEIA, HEIA and NCA.

This chapter continues on the theme of system classification by distinguishing between different kinds of technologies. Some technologies are useful to an individual, others to a group of farmers, some are inherently more sustainable than others, and some depend on use of more inputs while others can help to reduce dependency on external inputs. Some technologies can be called exploitative; others are regenerative, as explained in the following section.

The application of technology is not necessarily progress, but may be a response to a shortage situation (Wilkinson, 1973; Schiere and De Wit, 1995; Slingerland, 2000). For example, in EXPAGR with ample feed available there is no incentive to plant fodder unless the existing fodder base runs out. In that case the farmers start to feed straws for lack of grass (LEIA), or they start to grow fodder based on fertilizer or legume-nitrogen (NCA). Farmers will not be convinced to vaccinate or house the animals if there is no danger of disease or predators; attention to veterinary aspects is a necessity, not a luxury. Then again, one technology may serve several goals at once and it may also carry disadvantages.

Housing helps to protect the animals, to facilitate daily chores or to conserve feed. It also costs money and resources, it constrains the freedom of the animals and it may even negatively affect the health of the animals (if poorly designed). Similarly, there is hardly any technology that comes alone. The use of a milking machine requires the availability of skilled technicians and training of a farmer; the introduction of purebred pigs requires more elaborate housing, arrangements for supply of genetic material and availability of veterinary care. In some cases even the local food habits may have to change – for example, when slow growing pigs with much fat are replaced by fast growing pigs that have relatively lean meat and that use certain kinds of feed more efficiently.

Technology in general can be defined as a method to overcome a problem. A farmer getting tired of milking the cow will be happy if there is a labourer or a machine to do it instead. Farmers who get tired of hand weeding the crops will be happy to use a hoe or chemicals to make the job easier. Any intervention can be considered a technology – whether it is technical, management or policy related – but there is one condition. The cost of the technology should be paid by a combination of the extra return and the saved effort, a "calculation" that tends to be done differently among farmers themselves, scientists and/or policy-makers.

TYPES OF TECHNOLOGY

Several different technologies exist in relation to mixed farming and the following distinctions are made in this report:

- Input/mechanization-based versus management-based technology
- Accelerating versus defusing technology
- Exogenous versus indigenous technologies
- Technologies for national and local problems
- Technologies for individual farmers and for society
- Exploitative and regenerative technologies.

As in any other classification there is always another way to distinguish between types of technology. The purpose of this classification is to show major differences. The interventions discussed later will, in practice, always be a mixture of those mentioned above.

Input- and management-based technologies

Input- and mechanization-oriented technologies increase the output of a particular animal or farm by using more inputs (feed, fertilizer, pesticides), by using a machine to save labour, or by digging wells if there is not enough water. Management-based technologies are focused on trying to understand the farm as a combination of soil, plants or animals; the region as a combination of farms, people, mountains, underground water reservoirs, etc. Proper management can help to avoid losses where possible. For example:

- The cow when it stands up in the morning takes some time to stretch and to urinate/defecate. Knowing this, the farmer waits for fifteen minutes before letting the cow out so that the dung and urine are collected in the animal shed.
- When building the sheds, the housing can be oriented either to let the sun shine in unnecessarily and at the wrong moment, or the sun enters at times when it suits the management.
- Much effort can be spent on collection and preservation of animal manure but appropriate management is required to apply it at the appropriate moment to serve its purpose.
- Veterinarians can be called in to cure disease but proper management (hygiene, timely feeding) and a keen eye (noticing the onset of a disorder) can help to prevent the need for a cure.

Accelerating and defusing technologies

The distinction between accelerating and defusing technologies is difficult, but it is related to the earlier distinction between input and management oriented technologies, and to the exploitative-regenerative technologies. An example of an accelerating technology is when a cow is given a medicine that "forces" the animal to become pregnant even when the cow's body is too weak to conceive. By forcing it to conceive again the cow will collapse and even more inputs are required to get the cow in shape again. Liming of the soil or cultivation of *Brachiaria* or cassava on poor soils can also be considered as accelerating technologies. *Brachiaria* is a fodder grass and cassava is a tuber crop for human consumption, but they have in common that they both grow well (initially) on poor soils (Photo 13). However, they exhaust an already poor soil even more. The same process takes place when goats or cattle are left to graze already eroded hillsides. Liming, also called marling, was a well-known form of fertilization in Europe for many centuries. It was done, among other reasons, to release phosphorus from the soil-nutrient complex by adding calcium (the main component of lime). Ultimately it led to soils being depleted, as evident from this example of old British farmers' wisdom of several centuries ago:

> *"It brings the grounde to be starke nought, wherby the common people have a speache, that ground enriched with chalk makes a riche father and a beggerly sonne."*
> *(Based on Lord Ernle, 1961.)*

or in more common language:

> *"Lime and lime without manure, makes both land and farmer poor."*
> *(G. Montsma, personal communication, 1993.)*

Defusing technology as opposed to accelerating technology tends to consider the problem to a greater depth, to look for its root causes and see how the basic processes can be stalled or reversed. Defusing technologies are relevant in NCA and they tend to be approaches that fallow land, that rotate the crops rather than treating them with chemicals (as in HEIA) to enhance fertility or to combat weeds/disease, or that reduce erosion rather than living with the consequences of erosion (Photos 1 and 14). By accepting natural limitations the defusing technologies would use tolerant breeds such as N'dama or Baoulé in areas with sleeping sickness, where otherwise heavy control measures would be required. When a cow is weak a diffusing management practice would be to stop milking her a few weeks sooner to help her rebuild the reserves necessary for the next lactation. Ultimately, the diffusing management approach implies attention to stress signals from the environment. It aims to reduce or adapt consumption and to restore the (local) resource base rather than to aim beyond (local) carrying capacity (Scoones, 1996).

Indigenous and exogenous technologies

The distinction between indigenous and exogenous technology is again not strict but it serves to make a difference between solutions generated and collected over many generations by farmers themselves and solutions that come from outside. By stressing the existence of indigenous technology one unlocks a vast potential of local knowledge and creativity that helps to modify external technologies for local use. Farmers can often come up with cheap solutions that would not have been generated from outside. Technical solutions from outside can certainly be useful – a vaccine against disease, a new way of conserving feed, etc. Others, however, do

PHOTO 13
Accelerating technology: growing of *Brachiaria* on poor soil is successful in the first years but it leaves the soil more exhausted if it is not accompanied by additional measures to enhance soil fertility (Peru)

PHOTO 14
An example of defusing technologies is found where farmers grow grass on hills and contour ridges to counter erosion and to rebuild the local resource base (Nicaragua)

not fit the local conditions and it may be better to use a local and cheaper solution. Typical examples of indigenous knowledge are ethnoveterinary medicine, where women know the local herbs to cure or prevent disease in their goats, chickens or other animals, and knowledge about the best timing for ploughing or planting. Use of indigenous technologies can also be relevant for coping with variations in soils, herding rights, local business people, etc. Indigenous knowledge is no "cure for all", e.g. it has no or almost no ways to cope with infectious diseases and occurrence of new disorders that come with development. Still, the use of local knowledge for local problems has been shown to improve the uptake of development programmes substantially.

Technologies for national and/or local problems
Some problems in animal production are felt more at national than at local farmers' level. Rinderpest, foot-and-mouth disease (FMD), classical swine fever, etc. cause damage for the farmers, but they mostly threaten

the export licences of countries. In other words, many farmers will not feel motivated to do things that are not in their own direct interest. Large-scale programmes will have to take that into account. A more dramatic example is when governments want food to be cheap and plentiful for urban populations. This may require the production of cereal varieties that are not liked by the farmers themselves. It may even force governments to work against local traditions that, for example, consider milk as a gift from the gods that is not to be traded for profit. Improper handling of such tensions will not lead to effective development.

Technologies for individual farmers and for society
Cooperatives are useful forms of farmers' organizations for development, for example to supply inputs, to work on credits and savings, or to work on a watershed programme. However, not all technologies are useful for collective action. For example, a vaccination campaign almost certainly needs the organization of groups, whereas the treatment of a broken leg or a difficult birth in an animal is based on individual interests (even though a cooperative can play a role). Farmers' study groups are collective action but farmers' investment decisions and cropping/livestock keeping strategies are the decisions of the farmers and/or their extended families themselves.

Exploitative and regenerative technologies
This distinction resembles that under the section *Accelerating and defusing technologies*, but it treats the time horizon of the farmers and of farming. Planting of *Brachiaria* leads to rich yields but also to more exhausted soils. The planting of legumes or investments in soil regeneration give lower short-term yield but ensure a livelihood for the future generation. In particular, the mixed farming of the NCA mode puts heavy emphasis on regeneration by involving technologies such as the use of leys, recycling of nutrients, mutual adjustment of cropping/livestock components and erosion control through fodder crops.

MIXED FARMING AND THE COMMUNAL IDEOTYPE
One particular issue in the suitability of technologies for mixed systems is in the concept of the "communal" ideotype. This term was coined by Donald (1981) for wheat breeding and it emphasizes that the yield of an individual plant needs to be made subject to the yield of the entire plot. More practically, no farmer is interested in high individual plant (i.e. subsystem) yield if it does not

lead to the increase of total farm yields. The principle of the communal ideotype and choice of technologies for the design and choice of technologies in mixed crop-livestock farms can be illustrated in many ways:

- In fodder crops it has been shown that napier grass and leucaena can be grown alone or in combination, particularly for NCA conditions. Growing of only one crop gives higher yields per crop but a lower yield from the combination (Table 2).
- A study for LEIA conditions in India with different crop-animal combinations showed that the use of cows with individual productions of 10 litres/animal/day led to higher total farm outputs than those producing 16 litres/animal/day (Table 3). The high-producing dairy cows cannot be fed with fibrous crop residues such as straw, i.e. the goal of high milk yield per cow would reduce total farm income, thus losing an opportunity to use these resources.
- A study of crop-livestock combinations in Kenya shows an added issue, i.e. the optimum combination of animals and crops. The large farmer benefits more from grade cows (high yielding) and the small farmer is better off with crossbreeds (Table 4).

Field observations and common sense support the point that farmers adjust parts of their farms to achieve larger overall output. In many tropical and temperate contexts they use crossbred cows or medium-producing animals rather than purebreds. In drought-prone areas farmers choose grain crops that also yield good straw/fodder, even at the expense of some grain yield (Joshi, Doyle and Oosting, 1994; Seetharam, Subba Rao and Schiere, 1995).

CONCLUDING COMMENTS

As in all farming systems, there are many different technologies. The use and application of a particular technology and/or management strategy are responses to a stress situation. In addition, what works in one place may not work in another. Participation is often undervalued in identifying local problems and modifying exogenous problem-solving technology to local conditions. The term "participation" here should be understood in a broad sense to imply that policy-makers should listen to farmers and vice versa; that the needs of animals are seen in relation to the needs of the crop and the soil; and that the price of food in the city is related to what would be required to have and maintain strong rural communities for long-term food security. It means that the optimum production level of an individual component is established in relation to the yield of the total. This principle is called the communal ideo-

TABLE 2
Yields of napier (*Pennisetum purpureum*) and leucaena (*Leucaena leucocephala*) in pure and mixed stands

	Yield (tonnes/ha/yr)		
	Napier	Leucaena	Total
Napier alone	12	-	12
Leucaena alone	-	8	8
Napier + leucaena	10	6	16

Source: Based on Mureithi, Tayler and Thorpe (1995).

TABLE 3
Optimum crop combinations, herd size and production at different individual cow productions with or without treatment of stover, when the farmer also has access to a small fixed area of good quality fodder

Individual production (litres/day/cow)	Total production (litres/day/system)	Herd size (cows/farm)	Cotton (ha)	Total income from milk and crop (Rs/day/farm)
2.0	5.1	2.5	0	22.2
6.0	9.5	1.6	0	35.4
10.0	10.6	1.1	0.4	39.1
16.0	6.6	0.4	1.0	27.6

Notes: Total area is 1 ha, i.e. 0 ha cotton implies 1 ha of sorghum, 0.4 ha cotton implies 0.6 ha sorghum.
The cows used in this modelling are "tropical" cows; they are smaller than their temperate cousins. A milk yield of 10 litres for an animal of 350 kg is comparable with a yield of 20-25 litres for a temperate cow.
Source: Based on Patil, Rangnekar and Schiere (1993).

TABLE 4
Optimal farm crop areas as calculated with linear programming for an area with mixed crop-livestock systems in Kenya

	Large farm holding	Small farm holding
Gross margin (Kenyan shilling)	7 952	6 560
Land used (ha)	7.98	3.9
Grade cows (cow/farm)	7	-
Crossbred cows (cow/farm)	-	5
Coffee (ha)	0.06	-
Maize (ha)	0.28	0.52
Beans (ha)	1.03	1.34
Potato (ha)	0.03	0.04
Banana (ha)	-	0.15

Source: Based on Kidane (1984).

type and requires governments, research and education systems to refocus their attention from parts of the farm towards the whole, whether it be a plot, farm, region or community. Farmers have always known this and in mixed systems have often opted for a combination that gives medium yields, e.g. for crossbreds rather than for purebreds, for dual- or multiple-purpose animals and/or crops rather than for single-purpose species.

Chapter 4

Selection of animal species

Animals serve numerous functions in mixed farming besides providing products such as meat, milk, eggs, wool and hides. They also serve sociocultural functions, e.g. as a brideprice or as gifts and loans that strengthen social bonds. Quite often they are a form of saving, and sometimes they just serve as ceremonial animals or pets. More than 60 animal species are directly useful to humans, but most attention tends to be given to cattle, buffaloes, sheep, goats, pigs, horses, donkeys and poultry. There are also more unconventional animals such as llamas, yaks, guinea fowl, ducks, bees and pigeons that can adapt to many conditions. Often these unconventional animal species consist of small animals that have the advantage of fast reproduction, i.e. a herd or flock of these species is quickly replaced after a calamity such as drought, a flood or disease outbreak. Large unconventional livestock such as camels, llamas, alpacas, yaks, bantengs and deer are adapted to specific ecological niches, often in mixed systems.

UNCONVENTIONAL ANIMALS

The term "unconventional" can be confusing because the animals may not be common in general, but they can be quite "normal" in particular niches. For example, the yak and the two-humped camel have an undercoat which enables them to tolerate low temperatures and wide variations in temperature, while the one-humped camel can live in hot arid environments because it has an efficient water conservation mechanism, long limbs and a heat-reflecting coat. These rather unconventional animals can live in marginal areas to produce meat, manure and milk (camels and yaks) and fibres (camels, llamas, alpacas and yaks). In addition, the camel is used for draught and transport in dry areas, llama and alpaca for transport in the Andes, the yak as a riding and pack animal in mountainous central Asia and the banteng as a source of farm power in Southeast Asia (Photos 15 and 16). Multipurpose animals such as these are important to sustain economic activity in harsh environments and are generally

PHOTO 15
Llamas in the Bolivian mountains

COURTESY OF RIJK DE JONG

PHOTO 16
Bantengs pulling ploughs on paddy fields near Malang (East Java, Indonesia)

associated with mixed farming (based on Reijntjes, Haverkort and Waters-Bayer, 1992).

DAIRY ANIMALS

Milking must have started after people had associated with animals for other purposes, such as for meat production based on hunting, keeping of live animals for sacrifice, or even use of animals for draught or transport. Milk is an important product from animals in mixed systems, but not the only one. Moreover, the feed quality in many of the mixed systems does not allow high production levels. The major dairy animals are goats, sheep, cattle and buffaloes (Photos 17 and 18); each of these has a place in mixed crop-livestock systems and they share a digestive system that allows them to utilize coarse feeds like straws, grasses and tree leaves. The advantage of the small animals such as goats and sheep is that they are suitable for poor people, among others, to start from scratch. Where feed sup-

PHOTO 17
A dairy goat that produced over 2 litres of milk per day at its peak and is the pride of the owner (Sri Lanka)

PHOTO 18
Crossbred cow being milked by the daughter of the family, with a calf on the side to promote let-down of the milk – the cow was reared on fodder produced in a mixed system under coconuts (Sri Lanka)

TABLE 5

Milk production of tropical and commercial goats and tropical and western cows

Dairy	Litres/day	Number of days of lactation
Small tropical goats	0.5-1	50-100
Commercial goats	2-4	50-150
Small tropical cows	2-7	100-200
Large western cows	15-30	300-350

plies permit, small farmers have started with a goat, a sheep or even a chicken, and have gone on to build a herd and eventually to shift into large animals like cattle and buffaloes. Goats and sheep have different grazing behaviour than cattle and buffaloes, and a mix of these animals can serve to use the variation of feed on and around the farm better. Cattle and buffaloes also come in a variety of sizes and with different characteristics. Apart from the traditional attachment that people have to their traditional breeds, it is the bodyweight, orientation to milk production and tolerance to diseases that determine their suitability to a particular situation.

Different levels of milk production are given in Table 5. These levels are determined by body size, genetic background, farm management, health, feeding level, etc. An animal of 600 kg that is well fed and that belongs to a large breed can easily produce twice as much milk and meat as a small animal of 200-300 kg, simply because of the difference in body size. Goats and sheep produce much less milk and meat per animal than cattle but they also eat less, so roughly speaking one can say that the production per kilogram of feed is quite similar for small and large animals. Local tradition also determines the choice for a particular dairy breed based on the shape of the horns, the colour of the skin, the fat content of the milk (higher in buffaloes than in cows), the colour of the butterfat (from dark yellow to pure white), etc. Some types of milk are even believed to have special medicinal value, e.g. goat milk is generally thought to be good for asthma patients, and the finer distribution of the fat in goat milk makes it easier to digest.

ANIMALS FOR DRAUGHT AND TRANSPORT[1]

The number of animals used worldwide for work is estimated at 250 million and it may well be over 300 million. A wide variety of work animals exists, including cattle, buffaloes, donkeys, mules, horses, camels

[1] Partly based on FAO (1996), Starkey (1996) and Starkey, Mwenya and Stares (1992).

PHOTO 19
Bullock cart with straw (Sri Lanka)

and elephants. They provide the means by which millions of families make a living; they can also contribute to the establishment of ecologically and socially acceptable production systems. The use of work animals can reduce drudgery, it can intensify agricultural production and it can help to raise living standards throughout rural communities. Animals provide transport and mobility, they help in water lifting for irrigation, milling, logging, land levelling, road construction and local marketing (FAO, 1996; Starkey, 1996). Females are often incorrectly believed to be unsuitable for draught as they are lighter and can pull the plough only at the expense of some milk production. As feed becomes scarce, however, there is a trend towards the use of female animals, albeit with a drop in milk yield.

PHOTO 20
Bullock cart on a sugar-cane estate (Dominican Republic)

Animals to work the fields
The introduction of animal traction can increase or accelerate the exploitation of the natural resource base, particularly in EXPAGR. For example, when yield per unit area starts to drop, an animal-drawn implement can lead to an increase of the total land area under cultivation. As a result, the declining yields are masked for some time and farmers are not compelled to farm the same land more efficiently. Extra land preparation can also help to accelerate the degradation of soil organic matter, thus enhancing the impression that farm power through animals and traction can help to increase crop yields so as to keep pace with population growth (Photo 21). Alternative, more defusing uses of animal draught occur where animals are used to allow better weeding, more timely preparation, ridge building for more infiltration, etc. Expanded land use is associated with the use of animal power in West Africa where it was introduced only in the last decades. More intensive use of the land through use of animal draught is found in the

PHOTO 21
Animal draught in dryland farming in Africa: horse traction in Burkina Faso

age-old tradition of animals that pull the plough or that puddle the paddy field in the NCA oriented systems of South and Southeast Asia (Photo 22). Another "accelerating" aspect of animal traction could be the increase of women's workload in some societies. Often more land can be ploughed with animal traction, also where women do much of the weeding and harvesting. Moreover, the preparation of women's fields is almost always given low priority and it depends on the local arrangements whether husbands, brothers or fathers are willing to plough the women's fields.

Also at village level the social organization determines the adoption and usefulness of animal draught. The introduction of tractor power in some areas of the Pakistani Punjab made small farmers less dependent on the bullock pairs of large farmers, i.e. fields could be ploughed more timely. The fast adoption of animal traction in Africa was experienced in areas with relatively developed crop marketing systems, particularly for cotton and groundnuts, as in southern Mali, central Senegal and much of Zimbabwe. The cotton development organizations had a strong interest in removing the constraints that farmers could encounter in the use of animal traction. Therefore credit schemes and veterinary services were made available and loans could be repaid from the sales of the cotton or groundnuts.

During the 1960s and 1970s, it was thought that the increasing use of tractors as in Europe and North America would also take place in African countries. However, by the late 1970s there were higher oil prices, foreign exchange shortages and numerous failed tractor schemes, which suggested that rapid motorization was not viable or practicable in many of the African smallholder systems. Donors and governments, even in oil-rich countries such as Nigeria and Cameroon, started to see animal traction as a viable alternative with economic,

PHOTO 22
Use of animal draught in Sri Lanka to work the field – a rather unusual combination of a buffalo and a cow

PHOTO 23
The choice of implements can determine the efficacy of animal power – a head yoke such as used here can sometimes be replaced with a yoke that allows easier pulling (Peru)

BOX 9
ANIMAL TRACTION IN WEST AFRICA

Extension services in the Gambia introduced animal traction with ox ploughs between 1955 and 1975, but by 1988 more donkeys were being used than oxen. Donkeys are inexpensive animals that offer more timely cultivation. The use of donkeys for animal traction came from the neighbouring country Senegal and was informally adopted by the Gambian farmers. Two extension schemes, one formal and the other informal, helped to make animal traction a normal part of the farming system in a relatively short period of time.

In Mali, farmers often take three oxen to the field to plough instead of two. The reason is that the oxen are in poor condition at the end of the dry season and cannot work well due to fodder problems in the area. The three animals per team allow for rotational use. The animals plough for an hour or so, after which one is exchanged, thus reducing the adverse effects from low nutritional status. In the village most of the households own at least one traction unit. This is vital because the short and uncertain rainy season dictates the rhythm of fieldwork. Timely ploughing and seeding are crucial to achieve a good crop in the area.

(Based on Starkey, 1996)

BOX 10
COWS FOR ANIMAL TRACTION

Cows are increasingly being used for draught as so-called multipurpose animals. For example, 25 percent of the farmers in one village in Senegal use draught cows. Compared with oxen they produce less liveweight gain, but the production of calves and some milk compensates for this. Liveweight gain in young oxen can be used for profit by selling the animals after a few seasons while buying new young animals for the work. In some villages in northern Nigeria this has led to a kind of specialization where young bulls are bought, used for one season, fattened and then sold. The opposite is seen elsewhere, for example in Guinea where farmers keep the animals until they are old and lose weight. The animals are considered as friends and for the farmers it is unthinkable to sacrifice this friendship for financial gain.

(Based on Starkey, 1996)

environmental and social benefits. Also in Latin America animal traction is now promoted as being less expensive, more flexible and more environment friendly than tractor use. A campaign from Nicaragua shows that one tractor costs as much as 30 pairs of oxen that can do the work of three tractors, or 115 horses that equal 12 tractors. In Central America, there are potentially 160 000 pairs of oxen that can do the work of 16 000 70-hp tractors. This is a large potential saving on foreign exchange and, after a full working life, the animals can still provide meat for human consumption (Fomenta, Nicaragua, 1994).

Animals for transport

Transport of goods and people by animals is also a very ancient practice. In mountainous regions such as the Andes or the Himalayas the transport of agricultural produce is often only possible on the backs of animals because the roads are too narrow and inaccessible for motorized vehicles. Llamas are used to transport goods in the Andes over longer distances, just like dromedaries that transport goods on their backs through the Sahara. Market days in towns of rural West Africa (Mali, Senegal and Burkina Faso) give a formidable sight of rows of donkey carts. The donkeys are tethered nearby with a bundle of grass or hay, or allowed to graze on nearby wasteland. The number of carts used increased greatly in the 1980s and they are used throughout the year to transport the harvest from the field to the threshing point or to the granary, to transport the crop residues and the manure produced, to transport bricks or other goods. Such rows of donkey carts are paralleled by long rows of bullock carts transporting sugar cane to the sugar mills, whether on large estates or in smallholder cane cultivation around the cane mills in India. As is typical in mixed farming systems, these animals eat the cane tops and straw that is left over from the produce that they carry.

PHOTO 24
Community organization can also revolve around ploughing activities – an example of community ploughing (Peru)

PHOTO 25
Donkey carrying milk to the market (Sudan)

PHOTO 26
Horse used to transport milk from the farm to town (Peru)

PHOTO 27
Donkey cart transporting hay to town (Dilaba, Mali)

PHOTO 28
A Friesian horse pulling a carriage in the Netherlands – a case of
animals having a ceremonial function and cultural value in a
western society

PHOTO 29
The same ceremonial and prestige function for horses is encountered
in Burkina Faso

PHOTO 30
Bullock cart transporting goods from the village to the city (Sri Lanka)

PHOTO 31
A cart with straw carried from the rural areas to the city (Sri Lanka)

POULTRY

The term "poultry" refers to birds, a class of animals that
produce eggs, meat, dung, feathers, etc. It includes a
variety of species such as ducks, geese, chickens, song
birds, fighting cocks and turkeys (Photos 32 and 33).
They often provide a living for the family or an individual by scavenging the homestead (chickens and
turkeys) and by grazing on harvested rice fields.
Specialized forms of production are possible but they do
not play a large role in the mixed systems discussed in
this report. Several kinds of poultry function as "watchdogs" (geese), for rituals, as a social activity or for betting
(song birds and fighting cocks). The particular features
of these animals in mixed systems are that they are
small, they reproduce easily, they do not need large
investments and they thrive on kitchen waste, broken

grains, worms, snails or insects. Common problems are
that poultry, like other animals, are susceptible to disease, theft or to predators. They can also cause tension
between neighbours and/or within the family between
men and women, for example who feeds and eats the
chickens. Problems with disease can be overcome, as
listed elsewhere in this document; good housing over-
comes problems of predators and thieves; while the tension in the family and between neighbours can be a
stepping-stone for social development programmes.

Ducks and geese

Ducks and geese are basically waterfowl, but they can
do without water and, except in some rare cases, the
keeping of these birds does not develop into large-
scale, high-input production. Rare cases of specialized

PHOTO 32
Turkeys in rural Madagascar

PHOTO 33
Ducklings in the vegetation of a Vietnamese backyard

PHOTO 34
Foie gras, a roadside advertisement for this delicacy in the Vosges (France)

PHOTO 35
A flock of runner ducks on a homestead that are commonly used to graze the rice fields after harvest (Indonesia)

production of these animals are, for example, geese for "foie gras" (fat liver) or ducks for meat in countries such as France and China (Photo 34). More typical for mixed farming systems is duck keeping for eggs in harvested rice fields, such as in Java, where fallen grains, weeds, snails and worms provide free feed for large flocks of special runner ducks kept by individuals from the village (Photo 35). The occasional use of geese as "watchdogs" is worth mentioning as an indigenous way to fend off thieves.

Scavenging chickens

Common practice categorizes chicken development in terms of traditional, backyard, semi-commercial, commercial and industrial systems. This sequence, however, implies a progression from traditional to industrial. In other words, it represents an undervaluation of the small traditional production units and it does no justice to the role that scavenging chickens can play in a household economy of mixed farms, i.e. it disregards

the importance of the so-called traditional poultry, also called family or farmyard poultry (Photos 36 and 37). Indeed, many extensionists tend to approach family poultry with a narrow mindset. They focus on so-called improved housing, and avoidance of inbreeding and good veterinary care – issues that are more related to the keeping of confined than scavenging chickens on free range. For example, inbreeding is not a problem for family flocks where new blood is introduced by exchange or gifts of eggs and/or male birds.

Improvements can well be achieved within this kind of animal production provided that development agents approach the subject with a proper mindset. Birds can be kept, for example, as a flock of some adult chickens, one rooster and some pullets that are all sleeping in a low-cost night pen or even in a tree during the night. Feeders are not essential, but old tin cans or bamboo poles can be an inexpensive way to avoid wastage of feed. The chickens need to be supplied with

PHOTO 36
Scavenging chickens on grass in a homestead (Peru)

PHOTO 37
Chickens on a compound – a typical case of cheap chicken production on free feed (Botswana)

BOX 11

EGGS VERSUS PULLETS IN NICARAGUA – THE PROBLEM OF BROODINESS

Many women in Nicaragua prefer eggs above hatched chicks. For that reason they tie a broody hen with a rope for a few days until broodiness has gone. The hen will soon start laying again. This results in a production of about 100 eggs per local hen per year, of which only ten are used for hatching. The other eggs are used for consumption or sold for cash. Even a one-dollar weekly income from the sale of eggs is substantial in the mixed farming system of Nicaragua where such an amount represents almost a daily wage. It is particularly beneficial because of the low level, cheap inputs that are needed. The chickens eat low quality leftover feeds from the harvest and balance their feed intake from scavenging.

COURTESY OF ARND BASLER

PHOTO 38
A chicken pen on grass – one way to protect the chicks from predators at night on an organic farm in the Netherlands

---

additional water during the dry season, and nests can be made from leaves or clothes, often under baskets; it is essential that some place of refuge is provided where the hens can hide. Keeping chickens on pasture is also possible (Photo 38).

Many projects have tried to improve poultry production by using better feed, health care, housing and imported cocks. However, the more elaborate technologies and project interventions are rarely used for village chickens because of the expense involved and because of the difficulty even to find the necessary inputs. A flock of scavenging birds uses almost no inputs and it can still be a viable component in the mixed farm. Scavenging chickens can be fed with waste products and grains to supplement the weeds and

insects that are found in the backyard. Moreover, eggs from scavenging chickens in mixed farming systems are preferred to eggs from the commercial farms. In Nicaragua the eggs from the mixed farming system are called *huevos de amor* (love eggs) and people prefer them because of their yellow yolks and freshness. Higher prices are also often paid for scavenging chickens than for meat from chickens produced in intensive systems, mainly because of taste.

Chicken husbandry is often regarded as an activity for women, especially in the case of small flocks of scavenging birds. When there are higher cash returns, the husbands tend to become involved, leaving women only in charge of doing the work. Indeed, small-scale poultry production is liked by women and it can be an important way to increase their income. Rural women in tropical countries often use their chickens as a source of cash for the purchase of basic items such as salt, soap or school materials.

A major problem in chicken keeping is the occurrence of infectious disease. Scavenging chickens within mixed flocks do receive a certain natural immunity because viruses and/or antibodies are transferred from the elder chickens to the younger ones, which also suffer less because of lower stocking density. However, one typical "killer" is Newcastle disease, which can considerably minimize a flock if the chickens are not vaccinated. Vaccination against diseases such as Newcastle disease can be economic, practical and well adapted. Most mortality in mixed farms occurs amongst the young chicks (>50 percent). Also adult chickens die from disease. Predators such as birds, snakes and rats can also cause much damage. In one area of Nicaragua it was found that during one year 8 percent of the adult chickens had died – 1 percent caused by disease and 7 percent by predators. The farmer may know some local treatments but the effectiveness of those treatments is often not very high, particularly in the case of infectious disease. Given such conditions it is not surprising, perhaps, that people sometimes do not seem to mind if chickens die. Musole, a Zambian extension agent, once gave the opinion that it was not a problem that chickens died from Newcastle disease. He said, "they always die fat", implying a feast meal, thus taking an attitude that makes the best of things in life that one does not like. In other words, the mindset of farmers is also important in determining the output and health of a flock

Night pens for chickens that scavenge during the daytime need to meet only one condition to prevent predators like rats, snakes and thieves from coming in – they need to be cheap (Photo 39). In many countries chickens are not provided with a night pen. They sleep in a tree and farmers sometimes use a metal guard around the tree to prevent predators from climbing. They can also use a small shed on poles, near the homestead. Dogs that sleep nearby can protect the birds. Unfortunately, extensionists tend to regard the construction of a good pen as the first action in poultry husbandry (Photo 40). This can be wrong and expensive, particularly when no balanced feeds are provided.

Successful improvements in the keeping of scavenging chickens on mixed farms are based on the following approaches:

• Introduce commercial strains in the system, e.g. hybrid layers. A local hen can hatch the eggs of these layers and they can produce well after adaptation.

• Decrease mortality of young chicks where losses of 50 to 80 percent are reported, e.g. due to predators. The loss of a small chick of one or two weeks old may not seem a great loss but the young animals of "today" are the adult animals of the future. A common way to protect young birds is the use of moveable rearing pens that provide protection for the first four to six weeks, the most vulnerable period.

• Train people to improve egg hatching results. Often when all the eggs have been sold, a hen starts to get broody. Women then gather eggs from neighbours but those eggs are not always fresh and they give disappointing hatching results.

• Supply limestone or other calcium sources to the

PHOTO 39
A night house for poultry (Kenya)

PHOTO 40
A well-ventilated night house for poultry (Indonesia)

GOATS IN THE HIMALAYAS

There have been significant changes over the past two decades in the farming system in the Hindu Kush-Himalayas. The shift towards nuclear families is associated with fragmentation of landholdings. With more effective community protection of common property resources the land-based feed resources for livestock are becoming more difficult to obtain. Farmers in the area have reduced the number of large ruminants while increasing feed resources (privately planted trees, shrubs and ground grass) to sustain small ruminants. The practice of planting fodder on private land has increased in recent years because of decreased access to public land. Goats are easily managed because they feed on a wide range of fodder, grass and shrub species that are planted around the homestead. Because of their size, their feed requirements are also nominal compared to those of large ruminants such as buffaloes, while goats have an advantage in that they are small and can be tended by children.

(Based on: Tulachan and Neupane, 1999.)

good layers, particularly to the hybrid layers in scavenging systems that are only fed with grains.
- Start up village organizations for vaccination schemes for diseases like Newcastle disease, ideally combined with other small-scale social programmes that address the real needs of women and young adults.

GOATS

The largest concentrations of goats are found in Africa and on the Indian subcontinent but every continent has its own species and subspecies. Some are more suitable for meat, others for milk production, but goats in mixed systems are multipurpose animals. They produce meat, milk, offspring, skin and hair; they serve as a savings account and they provide readily available money when needed (Photos 41, 42 and 43). Goat meat is highly appreciated in countries where pig and/or cattle meat is taboo. In large parts of India the goat is the major meat supplier. In Mediterranean countries goats are kept for milk, which is consumed fresh and as yoghurt or cheese. The fat content tends to be higher than in cattle milk and the same is true for the milk protein content.

Goats are often accused of causing soil degradation and erosion. However, in reality the feeding behaviour of goats is not necessarily damaging; it is more likely

PHOTO 41
Goats in a zero-grazing system with straw bedding to capture nitrogen and to provide comfort on an organic farm in Belgium

PHOTO 42
Goats browsing on the side of a dust road (South Africa)

PHOTO 43
Goats eating from a simple, cheap feeder (an old tyre) (Nicaragua)

BOX 13
GOATS ON THE ADJA PLATEAU, BENIN, WEST AFRICA

As elsewhere, the presence of domestic animals is limited by disease though poultry, sheep, goats and, to a lesser extent, pigs are kept. The goats are of the West African dwarf type and trypanosomiasis resistant. They are kept as a kind of savings account and sold to solve problems in the family, such as sickness, to buy maize (the staple food) or to finance religious ceremonies. The keeping of small ruminants is an activity that is preferred by elderly people who can no longer work the fields but who use the animals to generate income for buying food. In some villages the goats are tethered during the cropping season to avoid crop damage. During this period the animals do not receive sufficient feed. Ongoing alley cropping trials with field-owning farmers could ease the fodder problems, by feeding tree leaves as fodder after a first pruning is used to fertilize the crop field.

An advantage of tethering the goats is that mortality rates are lower when compared with the completely free roaming system. The tethered animals are in closer contact with the livestock keepers and problems are noticed sooner than with free roaming animals. Health care is not well developed; although vaccination against small ruminant pest is possible through the extension service, the coverage is minimal. This has been blamed on the vaccination procedure of the extension service. A more active role of the villagers has improved the coverage, while traditional medicine has been used to cure diarrhoea with variable results.

(Based on Gbego, 1992; Gbego and Van den Broek, 1992.)

that they are left to graze after the land has already been deforested for cropping and timber or left vulnerable after overgrazing by cattle. Goats browse and prefer woody species and they normally find sufficient feed in 4-5 hours' grazing time, including walking time – contrary to cattle, which need double this time to satisfy their feeding needs. After droughts when cattle have died the restoration of herds starts with the small ruminants, because their reproductive cycle is short and their numbers can increase rapidly. The goats are then sold to re-obtain cattle from elsewhere. This is how agropastoralists in the Sahel restored their herds after the droughts in the 1970s and 1980s. The small size and relatively low individual values bring goats within the capacity of low-income farmers: "the goat is the poor man's cow". In many areas women and young adults also own goats, whereas cattle are almost exclu-

sively owned by men. To avoid damage to crops and to still take advantage of their benefits, goats may have to be kept in a zero-grazing system, often on slatted floors. The dung and the urine fall on the ground and in the process are mixed with poor quality feed leftovers used for compost while the goat is kept under hygienic conditions.

CONCLUDING COMMENTS

Many animal species are found on mixed farms, each of them occupying their own niche where they play a role in the functioning of the farm. They provide transport, draught, meat, milk and hides and act as a savings account to provide readily available cash when the need arises. There is an overlap in the functions and characteristics of some of the animal species, but they are all important for the running of the farm. Depending on local conditions, different types of animals are found. Not all species are found on the same farm. Llamas are found in the highlands of Latin America, buffaloes in Asia. Wealthy farmers tend to own large livestock and resource-poor families tend to keep poultry and small ruminants. Within the farm, men often own the large livestock such as cattle and camels, and the women and young adults own small ruminants or poultry.

Chapter 5

Traditional technologies for animal production

ANIMAL HEALTH

Animal health is essential in mixed farming systems, but veterinary health care knows failures as well as successes. The major successes are those of vaccination campaigns; the major failures occur where the veterinary services are largely unable to reach small farmers with their individual problems. Veterinary care in terms of prevention can help to improve livestock production. However, in traditional livestock systems the cattle owner mainly tries to avoid major damage at minimum cost and without depending on expensive and possibly erratic services from outside. Part of the health care therefore lies with local technicians using indigenous veterinary medicine, and learning to cope with disease – by spreading risks, by using animals tolerant to local diseases, and by running a low-cost operation.

Many diseases occur naturally and regularly depending on local ecological and sociocultural conditions, but changes in the world have also had their impact on local systems with consequences for animal health:

• War has forced people to leave their villages. By taking their animals with them, some diseases have spread to neighbouring areas or countries. For example, Mozambican refugees from the north spread African swine fever to peri-urban settlements around Maputo in the south.

• Population growth increases the pressure on land. Cattle and small ruminants have to move further from the village to less favourable pastures, which causes stress, and in addition there may be more contact with herds from other villages.

• Commercial movement of animals has led to the spread of disease. For example, rinderpest was introduced into Africa by a shipment of infected cattle from Asia in 1889 to feed Italian soldiers. It was catastrophic for cattle and wildlife. To replace these losses, in around 1903, cattle from Tanganyika were moved to Zimbabwe and this shipment spread theileriosis over southern Africa.

The actions that can be taken to cope with the effect of these changes are many, though not all problems can be solved on a small scale. One technological and futuristic approach lies in better mapping and prediction of animal diseases (J. Slingenbergh, pers. comm., 2000). Another lies in the continuation or modification of vaccination campaigns. Still another approach lies in the use of barefoot doctors and in recognition of the benefits (and sometimes constraints) of indigenous medicine. Some of these approaches relevant to mixed farming systems are given below.

Vector-borne diseases: traditional and modern treatments

Humid tropical regions harbour a range of endemic infectious animal diseases. Many of these diseases are caused by micro-organisms that hide for part of their lives in a tick, a mosquito or a fly (called the vector). Disease pressure is related to the density and the species of the vector. Cattle reared in infested areas for thousands of years have acquired a certain natural resistance but animals may still suffer and die when the infection pressure is high, or when other infections burden the immune system. One way to overcome or prevent these vector-borne diseases is to use insecticides in so-called cattle dips. This allows farmers to maintain animals with higher production levels because natural resistance tends to be at the expense of milk and meat output. However, breeds with natural resistance can be reared with a minimum of expense and they will continue to be important in sustainable livestock production in Africa.

The main disease vectors in cattle are the tsetse fly, various tick species and some mosquitoes. Over one-third of the African continent is infested by the tsetse fly, which transmits the micro-organism causing trypanosomiasis (sleeping sickness). Local cattle and goats have more genetic resistance to these micro-organisms than exotic breeds with higher milk or meat yields. The best-known tick-borne diseases of cattle are theileriosis, anaplasmosis, babesiosis and cowdriosis. African swine fever is a tick-borne disease of pigs to which exotic breeds are highly susceptible.

Vector-borne micro-organisms are hard to control but scientists and farmers continue to try with varying levels of success. A traditional way to protect livestock is to avoid areas of high risk in certain periods. For example, Fulani cattle keepers in the subhumid zone of West Africa avoid infested grazing areas, and they minimize the time spent at watering points where vectors are most likely to occur. In the wet season, grazing is delayed until late in the morning, as worm infestation on grass is highest early in the morning. Fires are built next to cattle pens to keep away biting insects at night and, when outbreaks do occur, the livestock owners avoid infected areas. Because of increased population pressure, however, it is increasingly difficult for nomadic cattle owners to avoid unfavourable areas, even by trekking long distances. The most economic approaches from formal research combine control and protection measures in a ratio that depends on socio-economic and epidemiological conditions.

Veterinary and animal health research uses genetic approaches in cattle, sheep and goats against vector-borne diseases. They work primarily on the genetic resistance of cattle breeds to the micro-organisms and to tick vectors, on genetic resistance of goat breeds to trypanosomiasis and on cattle breeds with natural resistance to ticks. In addition, genetic engineering of breeds with genes that favour resistance against diseases is a relatively new avenue in the battle against animal diseases and not yet proven to be effective. Another approach aims to reduce the vector burden, as has been done for decades, for example by the dipping of cattle against ticks. This is successful in the short term but it does not have a lasting impact on the tick population, nor on the disease. In addition, ticks develop resistance to acaricides. New acaricides will have to be developed continuously, to stay ahead of the diminishing sensitivity of the ticks. Resistance continues to develop against these products as well, i.e. new products need to be developed continuously to stay ahead. The FAO strategy on ticks and tick-borne diseases programmes aims at promoting integrated tick and tick-borne disease control methods that include immunization when applicable and increasing awareness about resistance of ticks to acaricides.

The development of vaccines against tick-borne diseases and trypanosomiasis is a major line of research. Important progress has been made in understanding the way in which micro-organisms attack the host. The vaccines so far developed all have disadvantages. East Coast fever vaccine, for instance, is expensive and exists off live, virulent strains. Animals vaccinated for East Coast fever may infect ticks long after vaccination

PHOTO 44
Screen traps against tsetse flies – a novel way to reduce the vector population (Zambia)

PHOTO 45
Cattle dip to control tick infestation. The centre of the building consists of a "swimming pool" containing the medicine in which the animals are submerged (Kenya). Technically these "dips" function well; however, the problem of their actual functioning lies in the socio-economic world and its functioning around the dip.

and contribute to new outbreaks in subsequent years. Immunization against some tick species has been tried and may have some possibilities. When this approach has effect, the tick population could be suppressed without damage to the environment.

Traditional, indigenous disease control and treatment
The interest in ethnoveterinary practices is growing. Many of these practices offer viable alternatives to conventional western-style veterinary medicine especially where the latter is unavailable, unaffordable or inappropriate. Ethnoveterinary medicine can provide low-cost health care for simple animal health issues though it tends to be ineffective against infectious diseases. Ethnoveterinary remedies are often based on knowledge and tradition from folk medicine for human use, such as in India on the *ayurveda* (the books about life). Most of the plants used are easily available but non-plant substances are also used. For example, warm stout is given to animals after they have given birth to help remove the afterbirth and cobwebs are used on cuts to help stop the

bleeding. Some of the plants used are multipurpose such as guava (*Psidium guajava*), bamboo (*Bambusa vulgaris*), rice (*Oryza sativa*), turmeric (*Curcuma longa*), aloe (*Aloe vera*), banana (*Musa spp.*) and *Kalanchoe pinnata*. These plants are either already found on farms or they can easily be grown. Many of these plants also have a food value. For example, an excess of green bananas can be ground, boiled and fed to stock as a source of carbohydrates and iron. Guava fruits and leaves contain useful vitamins. *Cymbopogon citratus* and *Ocimum gratissimum* can be used to make delicious teas. Medicinal plants to treat ruminants are used mainly for internal parasites, internal and external injuries and pregnancy-related conditions. Farmers usually boil the plants to make a decoction. Other plants are administered as teas, in which water is boiled and thrown on to the fresh leaves, which are left to steep (an infusion) and then administered once or over a period of days. Bamboo joints, thin-necked bottles or other appropriate instruments are used to drench the animals. As with any technology, care has to be taken in the use of indigenous medicine and application of knowledge. However, more attention to the potential of these approaches is likely to unlock a vast area of useful knowledge for conditions where modern medicine is out of reach.

HOUSING AND MANAGEMENT

Domesticated animals live, by definition, in and around the homesteads of people. Animals are either kept literally in the house, or in a pen or stable; they can be herded at shorter or larger distances from the home or they are only occasionally rounded up in a cattle crush or corral. Animals are not kept in sheds in conditions of EXPAGR, where land is abundant. Occasionally they are kept in corrals that are used for deticking of the animals, for marking, for vaccination or for castration (Photos 46, 47 and 48). As mentioned before, the EXPAGR mode is not associated with integrated mixed farms and therefore it is most interesting to discuss housing and management in LEIA and NCA systems.

Housing for animals demands investment and farmers must carefully consider whether it brings sufficient advantages to warrant such an investment. Many intermediate solutions exist, from grazing behind fences (Photos 49 and 50), to tethering cattle, to zero grazing where the feed is brought to the animals and where the dung is collected (Photo 52). Animals may be kept in stables year-round or only seasonally. During wet growing seasons animals tend to be kept on the homestead whereas afterwards they are allowed to graze on the harvested

PHOTO 46
Corral for animal management in Peruvian mixed systems based on EXPAGR. Large pastures are available and crops are grown without using animal dung

PHOTO 47
Marking of animals – an activity that is typical for grazing in EXPAGR, which needs a corral or management pen (Peru)

PHOTO 48
Castration of a male animal that is not to be used for reproduction. Animals are castrated to prevent unwanted reproduction, to make them easier to handle or to obtain a particular meat quality (Peru)

PHOTO 49
Cattle grazing in a LEIA situation in Kenya where pasture production starts to compete with maize production. Zero grazing based on crop residues and weeds offers an alternative

PHOTO 50
Intermediate conditions between EXPAGR and LEIA, where fences are erected to keep animals away from the crops (South Africa)

PHOTO 51
Damaged top sorghum caused by animals invading the crop field – an example of an EXPAGR system occurring side by side with more land strapped conditions (Nicaragua)

fields and in the bush. In semi-zero grazing systems, animals are kept in a stable or fenced enclosure for part of the day and particularly during the night, where they may be given some cut fodder. For the remainder of the day, they are allowed to graze.

The benefits, i.e. the reasons for housing and management, can vary widely between systems but they include aspects such as:

- Protection against climatic conditions
- Excreta management
- Disease control
- Prevention of theft or damage by predators
- Saving of labour
- Control of product quality
- Prevention of damage to crops

Animal housing in the warm tropics requires a shed design that keeps out rain and solar radiation, while allowing the free flow of air. Ventilation helps to cool the animals by allowing evaporation of water (sweat), thus cooling the air; by keeping the sun out, the place is also kept cool. In the hot tropics walls can be absent,

PHOTO 52
Cut-and-carry feed systems in LEIA – children fetch grass from roadsides in baskets (Indonesia)

although thick brick, stone or mud walls can help to lower temperature fluctuations between day and night. Roof overhang is important to keep out the sun and rainstorms when walls are absent. Trees provide shade and fresh air around the stable. They can be used to store feed (Photo 53) and roofing material should

PHOTO 53
A tree used to store straw – a very simple form of "housing" (Thailand)

PHOTO 54
Straw storage in an attic; the roofs are also made of straw, which is a cheap material that also keeps the inside cool (Sri Lanka)

PHOTO 55
Fodder fences with *Leucaena leucocephala* (Kenya)

PHOTO 56
Cultivation of oats as animal feed (Australia)

reflect the solar radiation and allow ventilation while preventing draught. To deflect solar radiation the roof can be painted white, made of reflecting materials, of tiles and/or plant materials that insulate (such as straw, grass or palm tree leaves) (Photo 54).

FEEDING TECHNOLOGIES
Different modes of feeding in mixed systems
Feed provides the energy and minerals that allow animals to stay alive, to grow, to produce and reproduce, etc. It can come as straw, grass, tree leaves, grains, tubers, insects, etc., depending on the agroclimatological conditions and on the mode of farming. Livestock in mixed systems of the EXPAGR mode depend mainly on grazing on wastelands, fallowed

croplands or distant grazing areas. As the systems move into the LEIA mode there is a greater need to use crop residues or industrial wastes. As systems move further into NCA there is a particular opportunity to use leaves from leguminous crops, or biomass from crops (green manure) that are grown between the main crops. Improved fallows, so-called leys, can use crops such as legumes or crucifereae (mustards) and even fodder grains, while living fences provide timber, fodder and posts (Photos 55 and 56). In HEIA systems it is common to grow special fodders or to buy feed from outside; mixed farms where animals and crops exchange resources "on the farm" are rare in HEIA.

Crop residues are relatively unimportant where there is plenty of land or other resources such as in EXPAGR and HEIA. However, use of crop residues for feed becomes vital in situations of LEIA and NCA where land becomes scarce. The straws from cereals, sometimes coarse such as maize, sorghum and pearl millet, and sometimes fine and slender such as wheat, oat or barley stover, thus form a major feed resource under many LEIA conditions. (See Singh and Schiere [1993; 1995] and Joshi, Doyle and Oosting [1994] for reviews and extensive literature references on this topic.) Unfortunately, the feeding value of such straw and stover is low. Cattle and buffaloes can eat little of them and the concentration of nutrients is low. As feeds become scarcer and of less value, it is better to give one feed in a less wasteful way (Photos 57 and 58) and/or introduce methods to overcome the nutritional limitations, e.g. through treatment with chemical or physical methods, through choice of other varieties and planting methods, etc.

Feeding of fibrous crop residues

Production of crops such as cereals and oilseeds yields two kinds of by-products that can be used as animal feed – the highly valuable grain and oilseed residues (such as brans and cakes – Photo 59) and the poor quality straws and stovers. The green fodders from legumes and specialized grass production that are part of the crop rotation occupy an intermediate position in terms of feed quality (Table 6). This chapter focuses on methods to use the fibrous crop residues for animal feed in crop-livestock systems (Photo 60).

Animals at low levels of production of milk, meat, draught, etc. do not generally require much supplementation to meet their nutrient requirements. They obtain sufficient nutrients by grazing on roadsides, or on fallow lands, or by consuming crop residues and kitchen waste

PHOTO 57
Goats walking on maize husks (Nicaragua). A wasteful method of feeding that causes large parts of the feed biomass to remain unused

PHOTO 58
Stall-feeding goats – the waste feed is put underneath the feeder to serve as compost when mixed with dung and urine (Kenya)

at home. During periods of scarcity they lose some bodyweight, which they can regain during the more favourable season. Low-quality feed does not provide sufficient nutrients to maintain the bodyweight of ruminant animals but it may be sufficient for survival. To achieve any production of milk, meat or draught over long periods, low-quality fodders must be supplemented

TABLE 6

Classification of crop residues according to crude protein content (CP), energy content (TDN) and CP:TDN ratio

Crop residue type	CP (%)	TDN (%)	CP:TDN
Category I: good quality			
Oilseed cake	28	70	0.40
Concentrate feed	15	65	0.23
Legume tree leaf	24	60	0.40
Category II: medium quality			
Medium quality grass	12	60	0.20
Rice bran	11	55	0.20
Mature grasses	10	55	0.18
Category III: poor quality			
Maize straw	6	50	0.12
Rice straw	4	45	0.09

PHOTO 59
Hydraulic press used to separate the oil from the seed for crops such as sunflower, soybean, coconut and cotton. The resulting residues (cake) have a high value as animal feed (Peru)

PHOTO 60
The straw from cereals has a low nutritive value in terms of digestible energy but it can be very valuable as a survival feed – here the straw has been made into a bundle for later use as feed (Nicaragua)

with better feeds and/or they must be treated with chemicals or other methods. Farmers commonly supplement low-quality fodder with concentrate feed ingredients such as brans, cakes or mixed commercial feeds. Treatments are less commonly used and they are useful only under specific conditions. Different treatment methods and supplementation are summarized in Table 7. Catalytic and strategic supplementation and urea-treated straw are treated in more detail.

Supplementation
Special feeds to supplement the lack of nutrients in straw-based rations can be used in several ways, called catalytic, strategic or substitutional supplementation. The latter is only applied in industrial systems where cheap concentrates/brans/cakes are available. Under these conditions straw is only fed as a source of fibre to aid digestion. The aim in these systems is to feed as much concentrate as possible, i.e. to substitute the roughage.

Catalytic supplementation
Supplementation of low-quality fodder with nutrients such as nitrogen and/or phosphorus is called catalytic supplementation if it is done to achieve improved rumen function. Small amounts of nutrients can result in increased intake and digestibility of straws. The supplement is said to have a positive associative or "catalytic" effect on digestion. The approach aims to increase the roughage utilization by using a minimum quantity of supplement. Obviously, this is attractive where low-quality roughage is cheap and where supplement is expensive, in combination with production objectives that aim at near maintenance growth. The most common example of cat-

TABLE 7
Feeding systems based on the use of fibrous crop residues

Feeding system	Description
Emergency and survival feeding	The use of any type of feed to achieve survival of the herd or animal, if necessary at the expense of liveweight and/or (re)production.
Catalytic supplementation	The use of small quantities of good-quality feed to improve digestion and intake of a basal ration of straw or mature grass.
Substitutional supplementation	The use of large quantities of supplement to supply sufficient nutrients for a desired level of animal output, if necessary at the expense of straw/grass intake, i.e. supplement substitutes the basal ration.
Straw treatment	Use of physical or chemical methods to increase straw feeding value or consumption
Chopping and soaking	Chopping implies the reduction of feed particle size, commonly at the size of a few centimetres or more, mostly to avoid waste of feed. Done alone or in combination with soaking.
Selective consumption (strategic supplementation)	Farmers and/or animals can select the good part of the feed, leaving the residue for animals of lower output or for other uses than feed.
Stripping	The use of leaves before they mature on plants for animal feed; mostly coarse grains such as maize and millets.
Thinning	The use of purposely dense sown plants for animal feed that are thinned and fed to the animals as the crop matures.
Use of variability	The term "use of variability" implies the use of differences in straw quality and quantity due to management, environment and genetic factors.
Adjusted cropping	A variation on the theme of variability (see above), i.e. crop choice is at least partly based on the nutrient requirements of the animals; or animals and crops are mutually adjusted, e.g. in the Flemish/Norfolk systems (see Chapter 7).

Source: Based on Schiere (1995).

alytic feeding is the use of lickblocks or other systems in which cattle can lick a mixture of (generally) urea and molasses (Photos 61 and 62). Lickblocks often contain urea as a supplementary element, as it needs to be fed slowly and distributed evenly. The block is a compact brick, placed before the animal to lick. This helps to meet the nitrogen requirement of rumen microflora, resulting in enhanced rumen activity and increased degradability of crop residues. Feeding of some kitchen waste or green leaves may achieve the same objective of maintenance.

Strategic supplementation and selective consumption
Another approach for efficient use of a limited stock of supplements is to use them only for certain animals: strategic supplementation. It is applied where better feeds are given to reproductive and young valuable animals, while adults or less valuable animals are left to lose weight. A typical but disguised form of strategic supplementation is the use of selective consumption (Photos 63 and 64). In this case the animals are offered a large amount of feed from which they can select the better parts such as the leaves while rejecting the rest such as the stems. The inferior feed refusals are then fed to less productive or idling animals or used for bedding. The final residue may be mixed with dung for composting or dung cakes. Strategic supplementation is applied frequently on farms when not enough better-quality fodder or feed are available for all the animals. Lactating cows, sick and weak animals or draught animals (both prior to and often during the working season) receive extra feed to perform their duties, to get well or do the fieldwork, respectively.

Treatment of straw
A range of treatment methods is available to increase digestibility and/or intake of straw. Simple methods include chopping or soaking of straw and more complicated ones involve steam treatment, e.g. for the treatment of sugar-cane bagasse (Rangnekar *et al.*, 1982). The most practical approach is based on the use of urea. Urea can be sprayed over the straw in a ratio of 2 percent with a 1:1 water:straw ratio, a form of catalytic supplementation. More relevant is the treatment of straw with urea where 4 kg of urea are spread with 50-100 litres of water on 100 kg of straw. The mix is kept in a heap for one to three weeks after which it can be used as feed with or without concentrate supplements. The treatment process increases the availability of energy from the fibres in the straw, apart from providing nitrogen for better rumen function (Photos 65 and 66). Experience suggests that the technology is most likely to work in the

PHOTO 61
Feeding lickblocks as supplements on grazing land (South Africa)

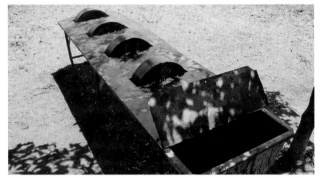
PHOTO 62
Feeding a liquid urea and molasses mixture by using roller drums (Nicaragua)

PHOTO 63
Selective consumption studied in digestion trials in Indonesia where sheep leave the central vein of the leaf

PHOTO 64
The practice of selective consumption in stall-feeding where animals leave the stems (Kenya)

PHOTO 65
Preparing urea-treated straw
(Sri Lanka)

PHOTO 66
Farmer feeding urea-treated straw (Thailand)

following situations (based on FAO, 1988; Singh and Schiere, 1995):

- when plenty of dry straw is available, free from fungal contamination;
- when farmers have slender straws from rice, wheat and barley rather than coarse straws from maize, sorghum or millets;
- when straw is cheap compared to other feeds and where there is a shortage of grasses or other green feed;
- when water is freely and conveniently available;
- when the price of urea is not prohibitive and where the cost of polythene covering materials is low;
- when labour availability is good, though small stacks do not require such high labour inputs at one time as the large stacks;
- when the animals are medium producers (milk or meat);
- when produce such as milk can be sold at a remunerative price.

Other straw feeding systems

A wide range of additional straw feeding methods exists, ranging from selective consumption, chopping and soaking of straw, breeding or management for better or more straw, to the use of adjusted cropping patterns and types of animals. The straw is chopped and/or soaked to increase the intake and to ensure that the animals eat the entire mixture rather than select only the better parts. Breeding and management for better or more straw relate to the selection of crop types and/or management methods that favour animal production. For example, plant breeders have long overlooked that farmers in conditions of unreliable rains prefer cereal varieties that will also yield good straw in the case of harvest failure (Photo 67). There are also varieties of sugar cane that produce good cane as well as fodder, a typical strategy for mixed farming conditions. Moreover, by densely planting the crops it is possible to get thinner stems, i.e. stems with higher digestibility. In a dense crop, a part of the crop can be removed before it matures, providing good feed and more space for the rest of the crop. Stripping of leaves is done to get green but maturing leaves of maize, cane and other coarse crops to be used as feed when their feeding quality is still good (Photo 68). Clearly, this requires a type of animal and production level that can use such feed; the concept of the "communal ideotype" illustrated with crop and animal choice (see Chapter 3). Whatever is done with straw, it is almost always a loss of nutrients and soil organic matter when straw is burned, even though that may be the easiest form of disposal (Photo 69).

Several methods of feeding and their usefulness for the farming system are indicated in Table 8. These depend on the mode of farming, the access to other feeds and the type and level of desired production. EXPAGR has an abundance of forest, bush and waste

land grazing, and feed resources that are a better source of nutrients than straw. Straw in these conditions is only useful to help the animals through a period of feed scarcity and selective consumption can be used. The use of straw for feed is more common in LEIA and NCA, systems that are both characterized by an adjustment of production objectives to resources. Particularly in LEIA, the shortage of feed, combined with the availability of labour and the adjustment of animal output to poor-quality feed resources, makes it relevant to chop or soak the straw in order to avoid wastage, or to make sure that a maximum number of animals is maintained. Straw use in NCA is determined by the need to recycle better or preserve excreta, to maintain soil structure and to avoid straw burning. Biomass from crop residues can also be needed for soil protection by mulching for direct improvement of soil. Often a combination of uses exists; even competition among straw uses can occur – as animal feed, for bedding to collect urine, for the soil as mulching, for the paper industry, for roofing or for fuel.

PHOTO 67
Better or more crop residues through improved management practices (India)

PHOTO 68
Woman collecting green maize leaves to feed the animals at the compound (Kenya)

PHOTO 69
Burning of straw, an accelerating approach used in EXPAGR systems where animals have enough access to other feeds. Burning is easy, but it leads to losses of nitrogen and organic matter (Sri Lanka)

Trees as fodder

In LEIA and HEIA there tends to be insufficient land to grow fodder as a special crop. One way to cope with a shortage of fodder is to use crop residues such as straw, as already discussed. Another approach is the use of leaves from trees that are specially grown on the boundaries of crop fields (Photos 55, 70 and 71). Use of trees for fodder can also be beneficial in EXPAGR where trees grow naturally in the bush or on the ranch. Feeding of tree leaves in such systems is a common strategy to overcome periods of feed shortage. Traditional feeding systems make maximum use of tree leaves, pods, seeds, etc. Farmers, but particularly the women, tend to be well aware of the habits and preferences of each animal (Photo 72). They know feed materials which are claimed to be beneficial for improving the quantity and quality of milk, and for improving the

TABLE 8
Usefulness of straw feeding methods per mode of farming in mixed crop-livestock systems

Mode of agriculture	Relevant feeding system
Expansion agriculture	- emergency feeding - selective consumption - catalytic supplementation
Low external input agriculture	- emergency feeding - chopping and/or soaking to avoid wastage - stripping/thinning - variability
New conservation agriculture	- straw treatment with urea - selective consumption - adjusted cropping - variability
High external input agriculture	- substitutional supplementation - straw as source of fibre in high concentrate rations

PHOTO 70
Leucaena, a well-known fodder tree

PHOTO 71
A live fence of fodder trees to keep pigs out of the garden (Peru)

PHOTO 72
Farmer transporting fodder leaves on his back to feed the animals at home (Sri Lanka)

PHOTO 73
Court bull with large horns that give aesthetic value (Sri Lanka)

PHOTO 74
Indigenous pig scavenging on the roadside (Peru)

health of their animals, etc. This is a typical case where farmers use indigenous knowledge to classify feed as very good, average or bad. They do so on the basis of the palatability and visible effects of the feeds on quality and quantity of milk, unlike researchers who look mainly at chemical analysis. The research and extension community can gain much if they learn to speak the language of the farming communities.

ANIMAL BREEDING AND GENETIC RESOURCES

Much concern exists worldwide about the need to maintain genetic diversity, on the one hand, and to increase the production of animals on the other. Whether goats, rabbits, chickens or cattle, many believe that the use of imported genes will increase the production of animals. Often it is even assumed that the import of genes is a miracle solution whereas, in fact, animals will only produce more if they are well taken care of, if they are healthy and well fed. Imported genes are not always improved genes even though there are conditions that call for import, often at the expense of local genetic diversity. Many traditional local breeds have already disappeared, together with their sometimes very rare characteristics such as horns, hair patterns and disease resistance, and influence on local folklore (Photo 73 and 74).

Fortunately it is possible to maintain the advantages of local breeds and also to reap the benefits from imported genes. Crossbreeding is the breeding of an exotic breed with a local one, aiming to maintain the best of both breeds. Where necessary, farmers and policy-makers can maintain local breeds. An example is the N'dama breed in the humid areas of western Africa that has particular tolerance to sleeping sickness. In other locations, particularly in cities with a good supply of feed, veterinarians, etc., it may be necessary to choose pure exotic breeds, whether chickens, pigs or dairy animals. Local breeds are a potential asset to countries for the adaptive traits that they have acquired over time, such as:

• tolerance or resistance to various diseases, including serious entero- and ecto-parasites;

• tolerance to fluctuations in feed resources and water supply;

• tolerance to extreme temperatures, humidity and other climatic factors.

BOX 20
CROSSBREDS IN UGANDA

In Kabarole (Uganda), the mild montane climate was favourable to European breeds of dairy cattle. In the 1950s and 1960s the government actively promoted dairy production, establishing a livestock-breeding centre in the district. This resulted in large numbers of local cattle being replaced by smaller herds of crossbred dairy cattle. Pastures were enclosed to control disease and to increase production, implying less common grazing land being available for local herds. Grazing areas were privatized and enclosed. The introduction of zero grazing and semi-zero grazing systems, in which the crossbred cattle spend most of the time in the stable, allowed farmers to collect manure and recycle nutrients more easily.

(Based on Walaga *et al.*, 2000.)

BOX 21
CROSSBRED CATTLE AND BUFFALOES IN THE HINDU KUSH-HIMALAYAS

The case of mixed farming in the Hindu Kush-Himalayas shows that access to milk markets, roads and veterinary services allows an increase in the number of crossbred cattle and buffaloes thus producing more milk. Buffaloes are particularly in demand because they are more resistant to prevailing diseases than exotic breeds of cattle and their milk has a higher fat percentage. Otherwise, the bulk of the livestock in remote areas of the Himalayan region consists of local species because these regions lack the necessary infrastructure.

(Based on Tulachan and Neupane, 1999.)

Farmers in areas where foreign "blood" was imported through development agencies are now starting to realize that their conditions are not always suitable for the "purebreds". Breeding for production levels that suit the local resource base is related to the issue of the communal ideotype (Chapter 3) and is often done by farmers when they "go back" from exotic to crossbred or even local animals.

Many livestock experts claim that environmental constraints should be removed before attention is paid to breeding because "only under favourable environmental conditions can animals express their genetic potential". However, genetic potential is not a single magic upper limit of the performance of animals. In the first place, production should not be measured only in the number of eggs or litres of milk. Particularly in mixed systems the production should be seen as a combination of dung, draught, meat, social satisfaction, etc. Secondly, the performance of genotypes may differ between environments and, thirdly, "genetic potential" also implies an animal's capacity to cope with harsh or unfavourable conditions. Therefore the genetic make-up of stock has to be judged in the environment in which they have to produce and genetic progress can be realized even under so-called suboptimal production conditions.

BOX 22
CATTLE BREED REQUIREMENTS

Preston (FAO, 1992c) worked out cattle breed requirements based on human nutrition demands. According to him the theoretical demands are 180 kg of milk and 50 kg of meat per year. The milk:beef ratio is 3.6:1. If less beef is consumed the ratio can be about 7:1. A dairy cow producing 4 500 kg of milk will produce, on average, a carcass weight of 250 kg per year. The milk:beef production ratio is 18:1. So, if milk production is based on a specialist system, either beef must be imported to make up the deficit, or there should be a parallel specialist beef production industry. In most countries this is a luxury operation that the farmers cannot afford and it is therefore necessary to develop appropriate multipurpose production systems based on integration of crops and livestock. In such systems, draught power and security are also important, i.e. the strategy must be one for optimization of a combination of production goals rather than maximization of a simple output.

(Based on FAO, 1992c.)

BOX 23
PIG SELECTION IN ASIA

Almost all domesticated pig breeds have evolved by natural selection from two major types: *Sus vittatus*, the wild pig of India, China and Southeast Asia, and *Sus scrofa*, the wild pig of Europe. The present-day so-called "improved" pig breeds arose from crosses between European landraces and oriental pigs. One author recognizes 87 breeds of domestic pigs in the world and a large number of varieties of pigs not recognized as breeds, but each having its own unique characteristics. Within the immense Chinese pig population, some 100 different breeds and varieties have been identified (Udo, 1999).

REARING OF YOUNG ANIMALS

It is one thing to obtain and use exogenous (or indigenous) breeds; it requires management to maintain healthy stock. The example of calf raising is taken here, but similar stories can be told of improvements through management in rearing goats, rabbits, chickens, etc. Many are based on hygiene, good use of critical inputs and housing, but most depend on the keen eye of the farmer. In mixed systems in particular a farmer has many things to keep in mind, and rearing of young stock can be different to that in specialized systems. Calf rearing may therefore not receive much attention and mortality can vary between 20 and 40 percent, or even higher in imported cattle. In farms with few animals, there tends to be competition between milk for sale and for consumption by humans and milk for the calf. In dairy-oriented systems the male calf has little value if the meat of young animals does not have an extra price, such as for white or pink veal. In cropping-oriented systems, on the other hand, the male calf as a future ox for draught has more value than female calves for replacement.

Climatic conditions, feeding and genotypes are important in the development of a calf over time into an adult animal. A hot climate gives lower weights at birth, a consequent lower growth rate, and lower liveweight at first service and calving. Poor feeding results in weak calves at and after weaning with high mortality rates, which in turn leads to little development of the mammary system, to a high age at first calving and to a low production of milk from the adult animal. Crossbreeding in particular results in a high variability of growth and corresponding liveweights and there is a wide range of weights at all ages at smallholder level for the combination of the above factors. Conditions can be improved by providing shade and natural and/or artificial cooling. Early feeding of hay and concentrate can correct poor milk feeding practices. The use of the adult weight of a particular crossbreed can help to determine the correct birth weight and growth rate one can expect. For example, a 300 kg crossbred cow is expected to produce an offspring weighing between 6 and 8 percent, i.e. 24-32 kg, with a daily growth of 300 g/d. From a 500 kg crossbred one expects a 30-40 kg calf with a growth rate of 500 g/d.

A well-reared calf is the productive cow of the future and various approaches have been tried to improve calf rearing. In dairy regions of the United States and western Europe there are calf-rearing clubs where farm children rear a calf in the best way they can while they register the costs of their exercise. An animal hus-bandry inspector regularly judges the technical and economic performance of the calves at the farm or a central place in the region while giving comments on the youngsters' efforts, and allocates prizes. Similarly, cooperatives in India, Sri Lanka and eastern Africa have held calf rallies based on the principle of judging at a central place.

Most dairy textbooks advocate artificial rearing to reduce the amount of milk consumed by the calf and leave more for family consumption and/or for sale. This approach is borrowed from conditions where artificial milk is cheaper than milk "direct from the cow". In many smallholder conditions the situation is different. Artificial milk is not easy to obtain, and restricted suckling is a good option. Restricted suckling is a practice in which a calf is allowed to suckle after the mother is milked; it reduces or prevents mastitis, a problem common in herds with artificially reared calves. It is

---

BOX 24
TRAINING IN CALF REARING
The National Dairy Development Programme in Kenya found a calf mortality of 20 percent in 1980 which, through training of the farmers, was reduced in four years to below 10 percent even though differences between male and female calves persisted (12 percent and 8 percent, respectively). Similarly, in smallholder dairy programmes in the United Republic of Tanzania (Kagera and Tanga regions), an emphasis on calf rearing reduced calf mortality to between 5 and 10 percent (De Jong, 1996).

BOX 25
TRAINING AND CHEST GIRTH MEASUREMENT
Sri Lankan farmers were trained in calf rearing through their cooperative society. They were given a measuring tape to record the chest girths of their calves and to compare them with a target growth line. A technician of the cooperative society regularly visited the farmers to check the weight of their female calves and to provide a small bonus if the calf had reached the targeted weight. Out of 1 850 calves registered within a six-month period some 916 had received a bonus at 18 months and 603 calvings were recorded at an average age of 30 months at the premises of the participating smallholders. The cost of the scheme amounted to US$ 100 per heifer that calved. The measurements also assisted farmers in the sale of their stock against buyers that tended to estimate live weights downwards (De Jong, 1996).

also very convenient for the cow in the case of poor milkers and in cases where the farmer has no time for milking, or when he/she wants to attend meetings away from the farm. Bucket feeding is more troublesome, especially with calves that carry zebu blood, because they cannot easily drink from the bucket. In addition, feeding with artificial milk powder can be troublesome due to low-quality powder (damaged by high temperature and humidity), difficulty in getting hot water and cleaning practices that are below standard.

PROCESSING OF MEAT AND MILK

Animal products such as milk and meat are difficult to keep over long periods. They can spoil and become unsuitable for human consumption in a matter of hours, particularly in hot and unhygienic conditions. A range of preservation techniques can be found to process this category of products, including heating, smoking, salting, fermenting (to produce lactic acid), drying or concentration to reduce the moisture content of the raw materials. For example, dried, smoked or salted meats are prepared to preserve the meat and to change the flavour and texture to increase variety in the diet. Conditions under which animals are slaughtered are often unhygienic and allow bacteria to be transmitted to meat by flies, animals and birds. Harmful microorganisms may grow on meat and cause food poisoning when eaten. These, together with infectious organisms such as parasites that grow in the meat, make careful selection of raw materials, proper handling, hygiene and preparation of meat products essential. In the case of milk, many traditional techniques can be found for producing indigenous products. Fermented milk products such as yoghurt and soured milk contain bacteria that aid digestion and that help prevent illness caused by other bacteria. In addition, fermentation removes milk sugar (lactose) from milk and facilitates digestion of the product.

Examples of rural processing of meat products include dried, salted and smoked meats. They are found widely in Africa where they are important traditional foods. Possibly the best known example is biltong, which is used as a snack food in southern Africa, but similar products are made in other parts of the continent. Biltong is made from strips of dried, salted meat, which are dark brown with a salty taste and a flexible, rubbery texture. Quanta is a similar dried and spiced beef product used as snack food in north Africa. Spices are mixed with fermented honey and about 120 grams of spice (composed of 2.5 percent salt, 1.5 percent black pepper, 10 percent

PHOTO 75
Clean rabbit cages in a backyard with special cages for kitten rearing (Guatemala)

PHOTO 76
Calf rearing: artificial feeding of a calf (Peru)

PHOTO 77
Rearing of young animals: duck eggs in a village hatchery (Indonesia)

spiced chilli) are added to each kilogram of meat and mixed in manually. The evenly spiced strips are hung on a string suspended in a well ventilated, dust-free area and left to hang for five to seven days until they break with light hand pressure or are crunchy upon chewing.

Examples of dairy products are soured milk, cheese and butter. Traditional sour milk is a thick clotted product similar to yoghurt but with a stronger flavour and a more acidic taste. It has a similar shelf life of three to eight days and is used as a drink or as an accompaniment to meals in some countries. Preservation is due to the production of lactic acid by naturally occurring lactic acid bacteria in the untreated milk. Various types of fresh cheeses can be made with simple equipment; a pan or container and some lemon can be good enough to prepare simple but tasty cheese. Traditional butter and ghee are made by stirring sour milk until the fat coagulates and separates into the solid butter (fat) and liquid buttermilk. The butter is collected by hand and washed with clean water two or three times before packing. This butter can be heated to evaporate the remaining water. The result is ghee, a highly valued cooking and frying ingredient, particularly in Asia.

CONCLUDING COMMENTS

Diseases, housing and feeding of animals in mixed farming systems have been discussed here with emphasis on low-input conditions. It has been shown that in certain circumstances it is better to trust and build on local systems and try to improve them than to introduce new technologies, including exotic animal breeds. Local animals are more adapted to the environment and exotic, usually more productive, breeds are often much more vulnerable and need more care, better feed and better hygiene than local ones, involving higher costs and a strongly organized system. The rearing of young animals can profit greatly from good hygiene and a well-balanced use of scarce inputs. Generally it is worthwhile to assess the local situation and to take that as a starting point for changes in the systems to achieve higher and more sustainable production of crop and animal products. The experience of local farmers is likely to be of considerable help in such cases and, by participating in the decision process, they will be more at ease with proposed changes or the introduction of new technologies. Much can be gained if research and administration learn to understand the thinking of the farmers, i.e. their priorities, their perceptions, their restrictions and their creativity.

Chapter 6

Crop-livestock technologies

The previous chapters discussed issues about animals and their products, or a particular technology that improves the output such as better feeding methods and better health services. This chapter focuses on the relationships among the components of animals, plants and soil. As with other chapters, only a selection of issues can be discussed.

INTEGRATING CROPS AND LIVESTOCK
Most livestock products in mixed farming systems are derived from animals that are fed on local resources such as pasture, crop residues, fodder trees and shrubs. Another feed resource consists of by-products from village industries such as oil extraction, grain milling and cane crushing. The exchange of these resources saves money and reduces waste by recycling products within the farming system. They create employment and they can contribute to soil texture and fertility while being an economic incentive for rearing multipurpose animals. As in the discussion of the other topics, relevance of technologies depends on the mode of farming. For example, in EXPAGR the animals, crops and soils are not strongly linked. Animals tend to graze elsewhere (outfields), they are a savings account and they may bring home some dung and young stock to pull the plough. However, in NCA the crops and livestock are integrated, as in LEIA, and they both serve the same goals – to keep the farming sustainable and pro-

ductive. Still, under poverty conditions in LEIA, there is a tendency for good-quality feed residues to be sold off the farm, thus leaving the animals with mainly fibrous feeds. Therefore animals in LEIA tend to be a way to convert fibrous feeds, crop residues, poor-quality grazing land and weeds into draught, dung and urine, often at the expense of the land (Photo 78). In HEIA there is hardly any attention to interactions among soil, plants and animals because limiting nutrients can be bought from outside and waste is ignored until it becomes a serious problem. HEIA exists not only in temperate and rich countries, but also in tropical and poor countries (Photo 79).

CROP RESIDUE AND SOIL MANAGEMENT
Livestock can enhance farm output by intensifying nutrient and energy cycles, particularly in LEIA and NCA. In LEIA this mainly implies a closer recycling of nutrients in straws (crop residues and fodder), dung and urine. In NCA there is scope to include leys, anti-erosion bunds and, for example, adjusted cropping patterns. NCA also has the added advantage that it can both import some nutrients in fertilizer and feed while it aims to keep the better-quality crop residues such as brans and cakes on the farm. Crop residues can be used for animal feed as well as for soil fertility and can also be sold. The farmers have to choose how to use these materials and, whatever they are used for, the choice

PHOTO 78
Gully erosion caused primarily by overgrazing (Senegal)

PHOTO 79
HEIA: cattle standing in a corral with plenty of dung. The animals are fed with local maize and cottonseed cake; they will be slaughtered and flown to the capital for consumption (Peru)

needs to be accompanied with keen management to make the best of these resources. The fibrous crop residues have a potential in soil conservation through mulching but even the oilseed cakes can be used to improve soils. Some terms related to animal-plant-soil interaction are explained in Table 9.

BOX 26
CROP RESIDUE MANAGEMENT AND SUSTAINABLE LAND USE IN BURKINA FASO

On the basis of current crop residue production, 54 and 98 percent of the present ruminant population can be maintained throughout the dry season in the sub-Sahelian and north Soudanian zones, respectively. There might even be scope to increase the contribution of crop residues to animal production in these areas by increasing availability of legume haulms and by allowing selective use of stover leaves. If the availability of haulms is 20 percent higher, and if it is combined with selective consumption (see Chapter 5), it might be possible to increase animal production by 40-50 percent. The capacity of animals to use feeds selectively requires collection and conservation that minimize the loss of nutritive value of high-quality residues (legume haulms). Collected and conserved under good conditions, the nutritive value of cowpea and groundnut haulms can be quite high (> 20 percent crude protein and 70 percent organic matter digestibility).

(Based on Savadogo, 2000.)

BOX 27
CROP RESIDUE USE IN ZOUNDWEOGO PROVINCE, BURKINA FASO

Farm surveys in Zoundweogo province, located in the north Soudanian zone, indicate that on average 3.0 and 0.9 tonnes of stover were stored out of the total of 11.5 tonnes of millet and sorghum stover and the 2.1 tonnes cowpea and groundnut haulms produced per farm. Ranges of 2-8 tonnes of stored cereal stovers and 0.5-4.0 tonnes of stored legume haulms are also reported, depending on rainfall. Availability of transport (donkey carts) allows storage of more high-quality feeds. The stored legume haulms are used for animal feeding, together with cereal stover. The remainder is used as building material, as a source of energy for cooking or it is sold. The crop residues remaining in the fields are usually grazed and only the non-edible portions serve as soil amendment.

(Based on Savadogo, 2000.)

DUNG AND URINE USE AND THE CYCLING OF NUTRIENTS

The use of dung and urine excreted by cattle, pigs, chickens and other animals for crops is traditionally raised as an argument in favour of keeping livestock. Dung and urine contain several nutrients such as nitrogen, phosphorus and potassium, and the solid fraction contains organic matter that is important to maintain soil structure and fertility. Occasionally it is even believed that dung and urine have medicinal qualities, but more commonly the dung is used as fuel (as dungcakes, through biogas), while it can also be used to plaster walls when mixed with clay. The belief in the medical value may have some justification, e.g. through the occurrence of antibiotics in dung and through the presence of lignin derivatives in dung that help to make it sticky or even to give it bactericidal properties.

The use of dung and urine for soil fertility is most cited. It is particularly important in LEIA and NCA, but unimportant in EXPAGR. The term dung "production" is misleading because animals excrete only nutrients that they first have to eat. Animals can carry nutrients from one place to another, they can increase the rate of their turnover or they can concentrate the nutrients from the outfield to the infield, i.e. they concentrate the fertility of a large land area on relatively small cropping areas around the villages. The "production" of dung in HEIA is a disadvantage rather than a blessing and elaborate systems are designed to dispose of dung rather than to use it.

Organic matter comes from manure, household waste and compost, or from incorporating weeds into the soil. The tasks involved in producing and using organic fertilizer are divided among different members of the household. Young people or labourers tend to collect the crop residues, to dig compost pits when needed and to transport the organic fertilizer to the field while the head of the family is responsible for managing its production. Women make organic fertilizer from household waste and keep some of it back for use on their own small plots of land. The major nutrients in fertilizer and dung are needed to make plants grow and produce:

- Nitrogen is essential for growth, for making the stems, leaves and roots; it is an important component of proteins.
- Phosphorus is needed for the plants to flower, to bear fruits and to develop strong roots.
- Potassium is important for tuber and fruit enlargement while it helps to maintain the healthy activity of all plant tissues.

TABLE 9

Terms and technologies associated with crop-livestock integration, soil fertility and animal nutrition

Fallow	Allowing the land to remain idle so that it can rebuild itself in terms of soil fertility, soil structure and soil life; or to ensure that disease and weed pressure will be reduced
Ley	A cultivated fallow planted with fodder crops that actively suppress weeds or disease, bind nitrogen, mobilize phosphorus, add soil organic matter or reduce runoff, etc.
Infield	The cropping area around villages or homesteads that is fertilized with excreta and with litter from the outfield
Outfield	The grazing lands, waste lands and forests around the village where animals are grazed or the faraway crop fields of a village
Catch crop	A crop that is planted to "absorb" nutrients in the soil that are released through weathering
Anti-erosion bund	A ridge planted (on the edge of a terrace) to avoid runoff
Straw	The stems and leaves of mature fine cereals such as rice, wheat, barley, oats and rye
Stover	The mature stems and leaves of coarse cereals such as maize, sorghums and millets
Grain:straw ratio	The ratio between grain and straw, i.e. a grain yield of 2 000 kg/ha associated with a grain:straw ratio of 1.5 results in a straw yield of 3 000 kg/ha
Harvest index	The proportion of the above ground biomass (in the case of cereals) that is found in the grain, i.e. a harvest index of 33 percent results in a ratio of 2 kg straw:1 kg grain
Excreta	A combination of dung and urine
Manure	Farmyard manure, excreta (the solid part) of animals that is to a lesser or greater degree mixed with straws and other leftovers
Soil organic matter	Organic matter originating from plants and/or dung that occurs in several degrees of decomposition in the soil, and in different degrees of solubility. It serves to enhance structure (on heavy soils) and water/nutrient holding capacity (on light soils)
Zero grazing	The keeping of animals in a shed, based on feed that is brought in from elsewhere (on-farm and/or off-farm)
Mulch	The mix of dry and/or green materials that can be used to cover the soil for prevention, for example, of erosion or excessive temperature effects
Compost	Organic matter (with or without animal excreta) that is decomposed on a heap or in a pit before applying it to the soil

Manure and nutrient cycling

The amount and quality of urine and dung produced depends on the type of animal, its size and the type of feed as well as on the management of the farmer. One way to calculate the amount of manure produced is:

- One animal of 250 kg live weight has a feed intake that averages 2.5 percent dry matter of its live weight. It therefore consumes 250*365*0.025 = 2 280 kg of dry matter. With an average digestibility of 55 percent, the animal will produce 0.45*2 280 = 1 026 kg of dung every year.
- Small ruminants weighing 25 kg and feeding 3.2 percent daily on average of their live weight, consume 25*365*0.032 = 292 kg dry matter. If their average digestibility of the feed is estimated at 60 percent it can be calculated that one small ruminant produces some 117 kg of dry matter faeces per year (Defoer *et al.*, 2000).

After the harvest a part of the crop residues tends to be freely accessible to all livestock including those from outside the village. The residues consist of straw that is sometimes left in the field or carried home and they contain stubble and ratoon (green regrowth of the crop) which can be grazed (Photo 101). Another part of the residue is fallen grain – in Asia this is used by ducks. Weeds can also be considered as crop residues and in wetter areas or seasons they can form an important feed resource. Sometimes they are cut and carried home during the cultivation period when access to grazing grounds is difficult for animals.

The nutrient content of manure and other organic fertilizers varies according to the quality of feed and the way it is stored and handled. Table 10 gives an indication of the concentration of the main nutrients in the manure of cattle and small ruminants and other sources of organic fertilizer. Dry matter content of manure also varies widely; in cows on lush pasture it can be less than 15 percent but in sheep on dry forage it can be higher than 50 percent.

The amount and proportion of nitrogen excreted depend on animal diet. The urine and solid dung of animals fed highly digestible diets with a lot of protein contain much more nitrogen and, therefore, are more susceptible to nitrogen losses than excreta from diets containing greater amounts of roughage. Much of the urine nitrogen is lost via ammonia volatilization. Where animal management tends towards increased stall-feeding, the composting of fresh manure will have to play a greater role in minimizing nutrient losses. Pits or heaps that capture feed refusals, manure and urine and household waste need to be designed to minimize nutrient losses. Low-cost appropriate implements to spread the compost at the appropriate time over large cultivated areas such as the Sahel are also needed (based on Powell and Williams, 1993).

TABLE 10
Nutrient contents of manure and other organic fertilizers (percentage)

Organic fertilizer	N	P	K	Dry matter
Cattle				
Fresh manure	1.4-2.8	0.5-1.01	0.5-0.6	15-25
Kraal (litter)	0.5-2.3	0.22-0.81	0.77-5.44	40-60
Kraal (no litter)	1.5-2.5	0.2-0.6	1.5-2.0	30-50
Goat and sheep				
Fresh manure	2.2-3.7	0.25-1.87	0.88-1.25	50-70
Fresh green manure	2.0-4.3	0.1-0.3	1.0-3.4	
Compost	0.3-0.9	0.07-0.17	0.14-1.3	
Household waste	0.2-0.9	0.05-0.5	0.1-2.1	
Ash from farm cooking	0.2-0.6	0.1-0.6	1.1-2.7	

Source: Based on a compilation by Defoer *et al.* (2000).

PHOTO 80
A woman fertilizing a crop field with great care and precision (Kenya)

PHOTO 81
Corrals on cropland to fertilize the fields with the dung of cattle that spend the night in the corral (Sahel, Mali)

In various systems draught oxen, donkeys, horses and small ruminants are kept overnight in pens in the compound throughout the year, and the manure they produce is transported to the fields during the dry season. Droppings of donkeys, sheep and goats are often added to the heap of household waste. However, droppings from small ruminants are sometimes used separately, for example on infields to manure certain spots in the millet crop. Manure from small ruminants takes longer to have an effect on crop yields than cow dung, but once it has started the effect lasts for several years. As this dung is kept the nitrogen losses can be reduced by mixing with low quality biomass, i.e. with straws, leftover feeds, dry leaves, etc. The greener leaves still contain much nitrogen themselves and they are less capable of capturing surplus nitrogen from urine and dung.

Household waste and compost
Household waste consists of partially decomposed waste from a variety of origins. It plays no part in fertility management where land is abundant. However, farmers are prepared to work hard to retain nutrients on their soil when land for cultivation becomes scarce and when food demands increase (LEIA and NCA). For example, farmers in Dilaba, Mali, said they started applying household waste in the 1970s after the drought, combining it with cow dung and the droppings of small ruminants. The system is now widely used in Dilaba and it is also starting to gain ground in the surrounding areas. The amount applied by different farmers depends largely on how much they are able to transport and on how much they can collect (Photos 82 to 85).

Composting is a process to break down the organic materials to make the nutrients in the biomass accessible

BOX 28
CORRALLING LIVESTOCK AT NIGHT IN MOBILE PENS
In Burkina Faso, central Mali, the Niger and central Nigeria, overnight kraaling of cattle herds is often used to fertilize the crop fields. Farmers hire Fulani herdsmen to spend the night on the crop fields during the dry season. The herdsmen are either paid in cash for this, or there are other arrangements such as the right to use water or the possibility to cultivate a plot for themselves. In certain areas of India farmers are even told to pay one year ahead to make sure that the herdsmen will come by again next year. Animals are kept on a small portion of the field and moved after several nights. In central Nigeria this was a plot size of 0.04 ha used to cultivate ginger, a major cash crop for the area. The drawback is the sometimes heavy weed development on the manured spots.

(Based on Powell, 1986.)

PHOTO 82
Household waste heap near the compound and fodder storage on the compound (Dilaba, Mali)

PHOTO 83
Compost pit filled with organic waste in the process of decomposition (Dilaba, Mali)

PHOTO 84
The transport of compost with a donkey cart. The compost is moved to the crop fields (Burkina Faso)

PHOTO 85
Empty compost pit and stover storage above the pit (Zimbabwe)

to plants. It also helps to make a fine, easy-to-handle material and, if well done, the composting kills weed seeds. Compost is normally produced from a mixture of all kinds of organic waste, including crop residues and household waste. The technology is very ancient and there are many different ways to compost organic materials. The most common method uses a moist heap with enough nitrogen for the breakdown of the organic material that is regularly turned to allow air and oxygen to enter the heap.

Nutrient losses and their prevention

Different kinds of nutrient losses can occur during crop cultivation and post harvest:

- Stray animals can damage the standing crop during crop cultivation and post-harvest period, in terms of losses of nitrogen, phosphorus and potassium and in terms of feeding value (Photo 51).
- Straw stubble that is left on the field can be trampled or soiled and incompletely eaten: careful harvest could save valuable nutrients.
- Leaving the crop residues in the field causes a loss of nutritive quality and sometimes leaching of nutrients through rains and degradation processes that involve fungi.
- Lush regrowth, in some cases after sudden rain, can be poisonous, such as the formation of prussic acid in sorghum ratoon.
- When the soil is bare for long periods nutrients can be lost due to erosion (wind and water) (see Photo 78).

Animals leave their droppings while they graze the stubble, so even farmers without livestock will receive

TABLE 11

Cereal stover removals and manure returns during crop residue grazing in West Africa

Location	Total dry matter (kg/ha)	Nitrogen (kg/ha)	Phosphorus (kg/ha)
	Stover removal		
Nigeria	2 470	24.5	3.9
Burkina Faso	1 570	14.8	4.0
Niger	2 135	16.8	1.8
	Manure returns		
Nigeria	27-272	0.3-1.7	0.1-0.3
Burkina Faso	600-1 600	7.5-20.0	1.5-4.0

Source: Powell and Williams (1993).

some on their fields. However, the quality of the dung is likely to deteriorate if left lying on the field unprotected for up to six months, even when it is stacked and stored in the open (Photo 86).

Grazing animals do not "produce" nutrients in dung and urine; actually they remove more biomass and nutrients from cropland than they return in the form of manure, as shown by studies in some West African countries (Table 11).

Activities to prevent or reduce nutrient losses are:
- Penning of animals with or without bedding (Photos 41, 75, 88 and 89).
- Storage of crop residues (Photos 54, 82 and 85).
- Mulching of the soil with residues unfit for animal feed.

BOX 29

COMPOSTING IN MALI IN THE SAHEL

French extension workers in Siguine (Mali) promoted cotton and introduced composting and the use of manure during the colonial period. Compost production was compulsory and done on a collective basis. Villagers used crop residues and cow dung to make compost and they also applied crop residues as bedding in the cattle pens, which were at that time communal. The produce was shared among the farms and was used on cotton and groundnut fields. When the French extension agent left in 1942 the farmers stopped producing compost. After more research, however, farmers in Siguine identified declining soil fertility as a constraint on production. The researchers proposed composting as a method of dealing with the problem and some farmers decided to start making compost, but this time of their own free will. They dig the pit in August when the soil is damp and start filling it with household waste, droppings of small ruminants and sometimes with maize residues and local hay. They take advantage of the last rains to start watering the materials. Once the pit has been dug it can be used for many years.

(Based on Dembele *et al.*, 2000.)

- Breeding or managing for crop varieties with better or more straws or for cultivars that have more ratoon.
- Covering of soil with so-called catch crops that retain leached nutrients to build their own organic matter that can be used for animal feed or mulch (Photos 87, 88 and 90).
- Harvesting of crop residues before they overmature, the so-called stripping as described in Chapter 5, section *Other straw feeding systems*, and as shown in Photo 68.

CROPPING PATTERNS, LIVESTOCK AND NUTRIENTS
The survival of farmers in mixed systems depends to a large extent on a proper adjustment of soils, plants and animals. Techniques suggested in this report include:
- better use of fallow;
- use of leys and catch crops.

These techniques can help in several ways either to reduce erosion, to add nitrogen, to enhance phosphorus flow rates, to pick up leached nutrients or to provide feed.

BOX 30

FEEDING LIVESTOCK FOR COMPOST PRODUCTION

The uplands of Java are continuously cropped and they support dense human populations. The use of manure-composts could be the key to sustaining soil productivity on the poorer soils. The intensity of cropping in these areas precludes grazing and the majority of livestock are therefore permanently housed in backyards and fed indigenous forages cut from field margins and roadsides. Cut-and-carry feeding is labour-intensive and the supply of forage is often the most expensive input to ruminant production. Surprisingly, farmers collect quantities of forage greatly in excess of the requirements of their livestock. Housing facilitates manure-compost production. Excess feeding increases returns to labour, because the nitrogen content of the forage selected and intake of digestible organic matter (DOM) increase and uneaten feed is used to produce manure-compost. Farmers collect uneaten feed in pits beneath their animals so that it combines with faeces and urine falling through the slatted floors to produce manure-compost. Manure-compost is ranked by farmers as one of the most important outputs from livestock production. In the upland regions of Java, 90 percent of the fertilizer used on smallholdings is manure-compost. It seems that livestock are used to produce high-quality compost and that their integration into Javanese agriculture is essential to the sustainability of some of the most intensive cropping cycles in the world.

(Based on Tanner *et al.*, 1995.)

PHOTO 86
Manure heap in the open air: nutrient losses here are high (Nicaragua)

PHOTO 87
Catch crop sown between a (harvested) maize crop, to protect the soil and produce fodder

PHOTO 88
Traditional pigeon tower for dung collection in Iran

PHOTO 89
Pigeon tower in the Netherlands. Such towers are cleaned regularly and the excreta can be stored and used as fertilizer

Fallow

Farmers in most EXPAGR regions tend to depend on fallow as the technique for restoring soil fertility. This is possible because there is plenty of agricultural land available and entire fields can be left fallow for long periods of ten years or more. The system resembles shifting cultivation and the location where crops are cultivated changes every couple of years. However, over the years, with increasing population and changing technology, the fallow period has become shorter and, in areas with a dense population, it has disappeared almost entirely.

Leys and catch crops

A ley is a fallow that is planted with crops such as grass or legumes to regenerate the soil more rapidly. The grass is used as fodder if animals are present on the farm, but similar systems are used when only mixing between crops takes place. In Mali, for example, farmers grow millet and cowpea on the same infield year after year. When they find that the fertility of part of the plot has fallen significantly, they change the association to pure stands of groundnuts or Bambara groundnuts. The following year they apply organic fertilizer and then start cultivating millet/cowpea again. Such a strategy is widely used in the Sahel. In other words, a ley is a crop rotation that includes animals. To some

---

BOX 31
TRADITIONAL FALLOW IN MALI, SOUTHERN SAHEL
In the past fallow was the main technique to restore soil fertility, because there was more than enough land available. Nowadays, this technique is still used to restore fertility for bush fields in Siguine (farming mode still close to EXPAGR) although fallow periods are becoming shorter. Farmers there use the presence of *Eragrostis* sp., *Zornia* sp. and crusting of soil as indicators of declining soil fertility – a case of indigenous technical knowledge (ITK). Previously, sandy soils were rarely used for more than three years and sandy-alluvial soils for not more than five years. However, with continuous applications of dung from cattle and small ruminants the village fields are now generally cultivated for 20 years or more. In Dilaba (mode of farming tending to LEIA) all cultivable land is in use as a result of rapid population growth, while the small remaining "fallow" area is used as pasture for draught oxen. In contrast with Siguine, fallow is not a means for managing soil fertility in Dilaba and the amount of manure used is considerably higher.

(Based on Dembele *et al.*, 2000.)

extent it is also the system used in the Gangetic plains and in the Nile delta, where berseem and/or mustard are rotation crops used exclusively for livestock feeding, either on-farm or off-farm (Photo 90). The system is being tried in northern Thailand (Gibson, 1987), and is also practised in the Krishna Valley of eastern India where a legume (pilipesare) is the main rotation crop. However, use of legumes is also common in the United States, in western Europe and in the so-called Mediterranean systems that even occur in Australia (see Chapter 8).

Towards sustainable land use

Integration of crops and livestock is often considered as a step towards sustainable agricultural production because of the associated intensified organic matter and nutrient cycling. Residues of the different crops represent the main on-farm source of organic matter and nutrients. This combines well with the presence of livestock since animals play a vital role as capital assets for security and as a means of saving, for cash income and in nutrient flows. Management of crop residues in such regions is closely related to their utilization in animal feeding.

An advantage of the integration of livestock and crops is the added value derived from crop residues (especially those of legumes) in terms of animal products and income. However, to maintain the system in the long term it is necessary that nutrients from external sources are added. In the past, this was covered to some extent through fallowing and manuring contracts with pastoralists, but the growing demand for feed by the increasing herd (from both arable farmers and pastoralists), the shrinking area of cropland per capita and declining crop yields dictate the scope for improvement through integration of crops and livestock. The extra input must come from inorganic fertilizer as well as from concentrates or both. Often, the price ratios of fertilizer and grains are not conducive to the utilization of fertilizers, and development of institutional and physical infrastructure for cost-effective use of fertilizers and concentrate is required to trigger sustainable land use in the region. Use of concentrate can be remunerative for farmers but credit facilities may stimulate intensification of livestock production and thereby increase availability of nutrients for crop land (based on Savadogo, 2000).

ENERGY, BIOGAS AND NUTRIENTS

Animals can also play a role in the provision of energy. Sometimes this role is very negative where livestock

PHOTO 90
The use of berseem as a nitrogen-binding catch crop and animal feed along the Nile delta. Animals are used to carry the feed home. The system is labour-intensive, and these fields are used for wheat and/or rice in other parts of the rotation

keeping contributes to deforestation in large parts of South America, Asia and Africa (Photo 91), but it can also be positive, such as by transforming plant energy into useful work or by providing dung used for fuel through dungcakes or biogas to replace charcoal (see Photos 92, 93 and 94). Wood and charcoal will continue for a long time to provide energy for household cooking and for rural industries but increasing energy demands and increasing pressure on natural resources call for measures such as a more efficient use of biomass, improved natural resource management, fuel switching from, for example, charcoal to kerosene or liquefied petroleum gas, and for the development of alternatives such as biogas. The principal uses of biogas are for household energy, such as cooking and lighting, although larger installations can produce sufficient gas to fuel engines, for example, for powering mills and water pumps.

Despite the merits of biogas technology, biodigesters have only been widespread in India (over 5 million installations) and in China (nearly 3 million installations). In these countries it is common to find biodigesters even in remote villages provided there is enough water. For example, an ambitious programme in Nepal is accelerating market development of small biodigesters. In many countries, some practical experience has now been gained with the dissemination of biodigesters but larger-scale dissemination has not yet become popular, despite the assumed economic, environmental and social benefits of the technology. The main reasons seem to be that the multiple benefits of biodigesters must be acknowledged by the end-users to convince them to invest in the technology, and that a

PHOTO 91
In South America large parts of the forest are cut to let cattle graze.
This often leads to degradation of the environment because of
improper management

PHOTO 92
Charcoal production: it is not only animals that cause erosion
(Senegal)

PHOTO 93
Biogas installation on a rural compound (Sri Lanka)

PHOTO 94
Dungcakes can be used as energy for cooking – this is often the case
if wood is hard to come by (India)

market-driven institutional infrastructure should be in place to facilitate large-scale dissemination of the technology. Much work has been done on mixed crop-livestock systems including the use of biogas, for example in the well-known CIPAV project in Colombia where pigs, sheep, sugar cane and biogas are just some of the components studied (FAO, 1992c).

Three designs can be distinguished among the small-scale and low-cost biodigesters:
- the floating-drum, also known as the "Indian design";
- the fixed-dome, known as the "Chinese design"; and
- the flexible-bag digester.

Roughly speaking, a well-constructed fixed-dome biodigester has the longest lifetime: 20 years and more. The floating-drum digester can have a comparable life-

time, but the recurrent costs are higher as the steel drums have to be replaced every five to ten years due to corrosion. The lifetime of the flexible-bag digester is hard to indicate, as some do not last more than a couple of months, while others function for several years.

Complex microbiological processes take place within a biodigester. In a common farm-type digester, the temperature ideally varies between 30 and 40 degrees Celsius. The process can be divided into three steps. During the hydrolysis phase, the complex molecules in the feedstock are broken down into smaller molecules. During the second phase, volatile fatty acids are formed. Finally, during the methanogenic phase, other bacteria use the acids to produce methane. The methanogenic bacteria can only do their work under anaerobic conditions (without oxygen), i.e. it is important for a biogas plant to be built of leak-tight materials. In practice, the three phases of the process take place simultaneously in a digester where the different bacteria work together in a symbiotic relationship.

Many different types of organic matter can be used to "feed" a biodigester, but to allow the bacteria to do their work the organic matter should be accessible to them. This means that pretreatment is sometimes necessary, e.g. through chopping and/or composting of crop wastes. The easiest feedstock to use is cattle dung, as it already contains the right bacteria and the vegetable matter has been broken down by passing through the guts (and teeth) of the cow. Human excrement and manure from chickens and pigs are also useful, but they do not contain the right bacteria and they need a starter in the form of, for example, some slurry from a working biodigester. Neither the feedstock nor the water should contain toxins such as antibiotics, detergents and disinfectants. Toxins cause the digester to "go sour", causing it to give off an unpleasant smell of acids and, therefore, the dung of cows that receive antibiotics should not be fed into the digester. Cultural aspects might influence the selection of feedstock. For example, Hindus in India usually accept cow dung for use in biodigesters, while pig dung and human excrement are often not accepted, particularly in Moslem regions.

Initially, biodigesters were promoted only for their energy production. Once again, the thinking about a technology in a mixed system focused on only one aspect. Unfortunately little attention has been paid to the excellent properties of the slurry as an organic fertilizer of crops and trees. In other words, slurry has the potential of supporting crop production and, in addition, the process kills or substantially reduces the amount of pathogenic germs and seeds of weeds present in the feedstock. Another advantage is that nutrients that are already present in the feedstock, such as nitrogen, potassium and phosphorus, are made available to crops more quickly and efficiently, i.e. the biodigester affects the nutrient dynamics of the system. Further, it reduces the odours given off by manure and compost. Emissions of greenhouse gases are also reduced, as the biodigester produces renewable energy (biogas) and because fossil fuel based chemical fertilizer is replaced by an organic fertilizer (the slurry). A major disadvantage of the slurry is that it is liquid, with a dry matter content of less than 10 percent. Transport to the fields is therefore often complicated and the use of the slurry is usually limited to fields near the digester. Larger installations in industrialized countries separate the slurry into a liquid and a solid fraction; the solid fraction can be applied on fields at a distance from the digester.

AGROFORESTRY AND SOIL FERTILITY

Alley cropping is an agroforestry system particularly for the humid or subhumid areas, developed at the International Institute of Tropical Agriculture (IITA), Nigeria, but based on earlier farmer experience (ITK). In this system food crops are grown in alleys formed by hedgerows of trees and shrubs, preferably legumes. In most tests and projects the fast-growing species _Leucaena leucocephala_ and _Gliricidia sepium_ (indigenous in Asia [Philippines]) have been used. The hedgerows are pruned regularly, thus providing fodder and fuelwood, and/or mulch. They are pruned to reduce shading for the crops sown in the alleys and the approach is considered to be an improved bush-fallow system permitting continuous cropping, among other reasons, because the tree is a giant "catch crop" that recycles nutrients leached to deeper soil levels. The alley cropping system can sustain or even improve soil fertility and crop production. The system, like any other mixed system, produces tree products such as firewood, stakes for construction wood and fodder for livestock. The latter is mostly used by cattle and small ruminants. Alley cropping is being developed with different tree species and crops, focusing on several uses of the by-products from the trees.

Alley cropping as such is not the solution to all soil fertility problems. Like other technologies it works

---

BOX 32
CALCULATIONS FOR THE FAMILY-SIZE FIXED-DOME BIODIGESTER

Assume that a small cow produces on average 10 kg of wet dung a day, equivalent to approximately 2 kg of dry matter. If cattle dung is used as the main feedstock for the biogas, the dung of approximately six head of cattle (diluted with water) will be sufficient for a small biodigester of 9 m^3. This digester would produce about 2 m^3 of biogas a day at an ambient temperature of 25°C, sufficient for the cooking needs of a family of around six people. At 30°C, the same digester would yield 3 m^3 of biogas a day, sufficient for the cooking and the lighting needs of the same family. In this case, the biogas would replace an average of 10 kg of fuelwood and 0.5 litres of kerosene per day, or roughly 4 000 kg of fuelwood and 200 litres of kerosene a year.

PHOTO 95
Agroforestry in Kenya: a field with *Leucaena leucocephala*, napier grass and, in the background, banana

PHOTO 96
Agroforestry in Benin: cassava cuttings planted in a field with pruned oil palm

only under certain climatic and economic conditions. Disadvantages are that it requires much labour and that feed cannot always be delivered on time. It also takes up space, i.e. it is again a reflection of the communal ideotype serving several goals and adjusting to other components in the system. A more intensive form of alley cropping is that of agroforestry in humid lowland tropics in a broad range of systems, some of which are traditional, such as multistorey gardens. Semi-arid areas have low rainfall and tree densities are therefore lower, i.e. windbreaks can serve as a variation on the theme of agroforestry, for example in the Niger with neem trees. In highland agriculture, together with grasses and shrubs, trees can be part of an erosion control system, as found in the Rwandan contour ploughing example that uses *Grevillea robusta* trees on the contours (based on Müller-Sämann and Cotschi, 1994).

CONCLUDING COMMENTS

The fascinating aspect of discussing crop-livestock technologies is that one has to think beyond the confines of either crops or livestock alone. In order to understand the issues at stake one has to know details about crops, animals, soils and people – and all the associated details. The technical issues concern first the intricacies of integration, i.e. the mutual adjustment over time and space of crops and animals (crop rotations, use of crop residues as feed and use of excreta for crops). Secondly, they stimulate the (re-)discovery of techniques that were traditionally in use in many societies, but that have been forgotten because of the emphasis on only crops or animals. In other words, techniques such as the use of fallow, leys and catch-crops that have been the basis of many sustainable

farming systems in the past. However, the emphasis on specialized forms of production based on high input use has made them disappear from curricula, research agendas and policy fora. The rediscovery of these technologies can combine with new research techniques to improve our understanding of when and where these approaches can be useful. Thirdly, this kind of work represents a very interesting and relevant area for testing and elaborating modern insights and concepts from system analysis, also for other sectors of society such as industry, habitat and tourism. Concepts to be elaborated could be the hierarchy theory, the effect of energy supply on issues of scale (between-farm mixing rather than on-farm mixing when oil prices decrease), and on issues of diversity (sustainability will increase if the system does not depend on one kind of activity alone). Some of these issues are discussed in the next chapter, by looking at the management of technology at several levels and by including the organization of the "human factor" – the difficulties as well as the opportunities.

Chapter 7

Management at farm, regional and policy level

Most of the discussions thus far have focused on technologies that range from feeding of crop residues, to recycling of dung and use of vaccines or dips to cure or prevent disease in animals. It has also been shown that a clear distinction between management and technological innovations could not be made (see Chapter 3). However, this chapter elaborates some issues that emphasize management and organization at animal, farm and regional level, such as:

- mutual adjustment of parts and actions on the farm;
- organization of village or region for effective development;
- adjustment of research and teaching in view of priorities in mixed farming.

A proper mix of indigenous and exogenous knowledge is likely to be effective in development of mixed systems – just as it is necessary to ally properly the interests of men, women, production and sales, etc. It will be seen that all these issues reflect the concept of the communal ideotype at farm, regional or even national levels and among disciplines.

THE COMMUNAL IDEOTYPE REVISITED AT THE FARM LEVEL

The essence of the communal ideotype is that parts of a farm need to be mutually adjusted to ensure maximum output of the total. This adjustment should take place not only between parts in space (i.e. the goat has to be tethered to avoid crops being damaged), but also in time (i.e. the yield objective of the present crop may not be at the expense of future yields; that would be a case of soil mining). For example:

- High-yielding cows cannot make use of fibrous crop residues; i.e. higher output of the ecosystem can be at the expense of the output from the whole farm.
- A high production of an animal in one year can exhaust the animal in such a way that productivity can be adversely affected over several years.
- Farmers reject new cereal varieties because they do not yield enough straw or straw of a good enough quality for animal feed.

- The excessive use and/or improper application of fertilizers may help to increase the yield of a fodder crop at the expense of the health of the ecosystem, i.e. ditches become polluted or lakes are overgrown with weeds.

All these issues are typical in mixed systems but, as previously mentioned, they also occur elsewhere in society, thus making the work in mixed crop-livestock systems an ideal model for other sectors of society.

THE COMMUNAL IDEOTYPE AT COMMUNITY LEVEL

At village or regional level there are also examples of development based on involvement and mutual adjustment of all partners.

Landcare and Integrated Catchment Management, Australia[2]

Landcare and Integrated Catchment Management (ICM) are two community participative organizations with a diverse membership in rural areas that includes primary producers, nature conservationists, government and industry. Land care in this sense is not a typical case of mixed crop-livestock farming, but it does represent a practical example of how issues in mixed farming systems translate into issues in development of society in general. Both Landcare and ICM involve all members of the community in the decision-making process for the management of catchments. Both these organizations mould together private use of land for agriculture with a philosophy that all land is held in trust for the use of future generations of human and other life. They do this by planning for long-term, sustainable use of land at a catchment level and use public and private funds to implement the plans at a property level, the epitome of the communal ideotype in space and time. Members of the organizations work together, and work with government bodies, to share information about best practices and to make incremental steps towards sustainable land use. Nature conservation groups are part of both organizations and they contribute to the thinking about what is current best prac-

[2] Based on Roberts and Coutts (1997) and Campbell (1996).

tice for the needs of wildlife. They also help to implement the plans using both public funds and private donations of cash or kind. Landcare and ICM have developed into strong organizations that combine the interests of different system levels in space and time.

The Sukhomajri project in India[3]

Sukhomajri lies in the hills, not far from Chandigarh in northwest India. The hills, once heavily wooded, have suffered from increasing human and livestock populations, leading to overgrazing and severe erosion. Barely 5 percent of the uplands had vegetation cover in the 1970s, and erosion rates of 150-200 tonnes/ha were not uncommon. In 1958 a dam was constructed, creating a lake to serve the city of Chandigarh, but by 1974 over 60 percent of the lake was filled with sediment. To protect the lake, the authorities first tried to persuade the *gujjar* herdsmen of the village to stop using the hills for their cattle and goats, but with little result. The breakthrough came when it was decided to build a small earthen dam in the hills to provide water for the village itself and then to stabilize the catchment of this dam. The stored water was used to irrigate nearby fields and farmers were provided with subsidized seed and fertilizer. Yields were greatly increased but farmers who did not benefit continued to use the hills for grazing. It was then that the villagers collectively proposed that more small dams should be created to extend the irrigation system. They also suggested the creation of a water users' society, based on the principle of equity, to manage the water. A "coupon" system was introduced and families with little or no land could thus sell their water rights or use the water to sharecrop on land belonging to others who were short of water. Any member whose livestock was found grazing in the hills lost his or her rights to the water. The society was given responsibility for maintaining the dams and their catchments, for distributing the water, and for maintaining records.

From then on the village began to develop rapidly. The villagers sold off their goats and replaced them with high-yielding buffaloes to provide milk for the growing towns nearby. The buffaloes were stall-fed, using the rapidly growing fodder grasses in the hills. These first steps immediately brought benefits and helped enlist villagers for further experiments in improvement. For instance, the farmers began to experiment with green manures, which had not been tried before, and they developed new technologies

based on velvet bean and lablab bean. As maize yields increased and subsistence was assured, the farmers began to turn to vegetable growing. The project is on a self-help basis, with no subsidies provided. By 1990 the farmers, with their own labour, had constructed some 300 km of erosion works. Out-migration has largely ceased and the landless in the area are benefiting from an increase in the daily wage. Many landless have begun to establish rights to land that previously they had considered useless but which under the new technologies is proving productive.

SIWAA, the dry forest, Mali[4]

Natural resources are in short supply in the Koutiala area, one of the oldest cotton production areas of Mali. Villagers complain about townspeople cutting firewood and grazing herds in the bush. In an attempt to regulate the use of the natural resource base a pilot project with six villages has been set up. They have been collaborating with other users of the natural resources and with research, extension and local authorities to manage their territory. Over the years plans have been put to work to control grazing, cutting of firewood, livestock feed and other uses of the resources. The plans included intensification of agriculture, integration of agriculture and livestock, fodder crop production, erosion control, establishment of tree nurseries, reforestation, and adaptation of improved cooking stoves. Respect of traditional laws, for example on when and how to exploit certain tree species, has been included in the protocol. The whole approach has been based on participatory methods. Chosen villagers represented the villages and they

[4] Based on Hilhorst and Coulibaly (1998) and Joldersma *et al.* (1994).

PHOTO 97
Unrestricted grazing can cause damage to existing vegetation and seedlings of trees and grasses, leading to overgrazing, mining and deterioration of the resources (India)

[3] Based on Conway and Barbier (1990).

gave feedback to the village on results reached or measures proposed to control their environment. It took a long time to get the different stakeholders together. The process was slow, with conflicting interests, but in the end an agreement was reached, the protocol was tested and is now being implemented. The cutting of firewood is regulated. There are rules for grazing, rotational grazing and bush fires, and cutting heights of trees are fixed to enable regrowth.

Perceptions of the role of community forestry [5]

Farmers in the Hindu Kush-Himalayas had mixed reactions regarding community forestry. Some felt that it had a positive impact on fodder supply because of natural growth as a result of protection, but others thought that controlled harvesting of natural grass and leaves would decrease the availability of fodder resources. A minimum fee could be paid on a monthly basis to acquire fodder grass from community owned lands, but poor farmers did not find this acceptable. With the increasing scarcity of fodder, farmers searched for alternatives. A greater number of farmers addressed fodder scarcity through use of their own lands. In some villages, about 30-40 percent of the land-rich farmers have private land, *Kharbaris* (marginal sloping land), on which grass and fodder trees are grown more intensively. Increased fodder shortages and the high cost of feed provided incentives for the increased cultivation of trees on farms. There has also been an increase of growing leguminous fodder trees on the bunds of cultivated terraces. During the winter months, maize stalks and *badhar* (*Artocarpus lakoocha*) are fed to livestock. Farmers also save dry fodder (rice and wheat straw, maize, stover, etc.) during this period to feed large ruminants but, as the amount available is not sufficient to feed them all winter, in some villages farmers have adapted to winter fodder shortages by cultivating exotic oats, sometimes cultivated after the rice harvest, to replace wheat. Also, some resource-poor farmers have responded to the lack of fodder from their private land by making arrangements to harness fodder from the private land of resource-rich farmers.

POLICY MEASURES

Not only farmers and their communities have to change their attitudes, also the people and institutions at government level may have to collaborate in adjusting their priorities to make livestock development and mixed farming a success. Typical cases range from the privatization of veterinary care to the reorientation of research and extension programmes.

Veterinary services: privatization of veterinary care

Traditionally the tasks of veterinary services in developing countries include active veterinary care: vaccinations, health care and marketing of veterinary and feed products. However, state-organized veterinary services have been increasingly incapable of responding to the growing demand for veterinary care because of budget restrictions, inefficient services and inefficient handling of revolving funds. Therefore, the privatization of veterinarian services was stimulated to continue offering animal health care after the Structural Adjustment Programs of the World Bank and IMF. Multilateral donors, such as the European Union, support private veterinarians through credit schemes and trade guarantees and the functioning of veterinary care on an open market is expected to increase availability and to improve the quality of animal health care. Better adjustment of the services offered and services needed is expected from privatization, though this may work better for one situation than for another. The prevention of diseases of national importance may be less effectively tackled if farmers determine demand for services only.

Privatization has indeed been successful to some extent. However, the private veterinarian hardly covers remote areas because costs of transport are prohibitive and demand for services is low. In many African countries the adjustment process is progressing slowly, but the privatization of vaccination campaigns against bovine pest and contagious bovine pleuropneumonia has been quite successful in Chad (Domenech, 1994). The trade of veterinary products has been successfully liberalized in Mali, but at first the availability of private agents restricted the range of vaccination programmes and prevention against trypanosomiasis. The success of privatization also depends on the organizational capacity of the veterinarians and their willingness to work in an insecure financial situation. Donors have stimulated governments to change the legal framework to facilitate the work of these veterinarians. However, the application of this framework depends on the veterinary services. Veterinarians have professional associations that can put pressure on government to either activate or discourage the new legal frameworks and adjustment processes.

Privatization of the veterinary services in the Compagnie malienne pour le développement des textiles (CMDT) in the region of southern Mali has given

---

[5] Based on Tulachan and Neupane (1999).

mixed results. In some parts of south Mali, such as Kadiolo, the introduction of cotton and the related animal traction on the farm has been relatively recent. Veterinary services such as vaccination and deworming of oxen were carried out twice a year on the initiative of the cotton company (CMDT). Farmers were not fully informed about the usefulness of these programmes. Sometimes costs were deducted from the cotton yield without involving farmers, who were not likely to call upon the veterinarians themselves because they were unaware of the advantages of preventive treatment. At the same time, private veterinarians were not stationed in these regions, as the demand for their services was low. In these cases privatization had negative consequences because some farmers lost their only oxen that were acquired on credit. Also, in 1995 and 1996 when privatization was under way and when state agents and the CMDT were prohibited to intervene, an outburst of pasteurellosis occurred, while trypanosomiasis caused mortality in cattle. The cotton production of these farmers decreased because of the poor health of their oxen, and so did their capacity to reimburse the credit they had received for cotton farming and the animal traction equipment. In regions such as Koutiala, where introduction of animal traction started much earlier, farmers learned about the advantages of vaccination by experience as they invested their savings from cotton production in cattle. They were used to calling upon veterinary services for their animals. By experience they learned that only vaccination could prevent cattle from getting bovine pest, pasteurellosis, anthrax (endemic in the region) and contagious bovine pleuropneumonia. After privatiza-

tion those farmers did not hesitate to bring their oxen for vaccination to private veterinarians that were already stationed and equipped with drugstores.

Cotton by-product policy in Mali

A major by-product from cotton is the cottonseed, which is pressed to extract oil. The resulting cake is a valuable feed for animals. The demand on the world market is high and industries are keen to sell large quantities to make a profit. Exports help to bring in foreign currency but reduce the availability of the by-products for the farmers that have produced the cotton. This is a form of disintegration, i.e. export leads to specialized livestock production in HEIA systems elsewhere while exchange of resources on the farm (the ideal of NCA) is reduced. In 1994, however, the newly created Union of Cotton Producers in Mali demanded the by-products of "their own" cotton. However, government and industry argued that the pastoralists in the north of the country also needed these supplements. Their interest was to reserve a part of the industrial by-product for the people in the north because these mobile nomads and transhumant livestock holders move south in the case of feed shortage. Their migration to pasture grounds and forest areas that were traditionally reserved for others had caused ethnic tensions on various occasions.

Research and education for the development of mixed systems

Much of the modern teaching and research tends to be focused on single commodities and disciplines. This monofocused approach on either breed, or feed, or

PHOTO 98
King grass planted on farmers' oil palm plantation areas in Kalimantan. King grass serves as fodder for the farm animals

COURTESY OF B.N. UTOMO

PHOTO 99
Palm kernels can provide a useful feed. Here a farmer has fed the solid palm kernel waste to his animals, which they obviously enjoy

COURTESY OF B.N. UTOMO

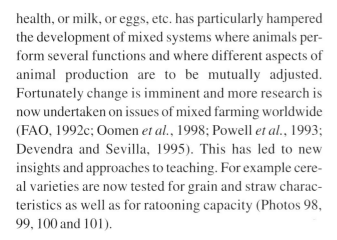

PHOTO 100
Testing of different rice varieties for characteristics such as grain yield, quality and quantity of straw (Sri Lanka)

PHOTO 101
Ratoon of a rice crop, i.e. young regrowth after harvest (Pakistan)

health, or milk, or eggs, etc. has particularly hampered the development of mixed systems where animals perform several functions and where different aspects of animal production are to be mutually adjusted. Fortunately change is imminent and more research is now undertaken on issues of mixed farming worldwide (FAO, 1992c; Oomen *et al.*, 1998; Powell *et al.*, 1993; Devendra and Sevilla, 1995). This has led to new insights and approaches to teaching. For example cereal varieties are now tested for grain and straw characteristics as well as for ratooning capacity (Photos 98, 99, 100 and 101).

CONCLUDING COMMENTS

Mixed systems that are based on exchange of resources need to consist of parts that are finely tuned to each other – the principle of the communal ideotype. The principle is valid at all levels in the system hierarchy, i.e. a herd is composed of animals that function as a whole, and a mixed farm adjusts to humans, livestock and the soil. A region should consist of mutually adjusted production systems. At a still different level, it is also necessary to adjust government systems and farmer preferences, as well as to adjust research and teaching methods so as to provide maximum benefit.

Chapter 8
Successful mixed systems

Sustainable systems are those that have existed for a considerable time. However, it must be recognized that no system can exist forever, i.e. any system becomes unsustainable sooner or later. All systems have to adjust continuously, particularly when the relationship between the system and its resource base becomes seriously out of balance. An example of a system and its resource base is the case of the infield/outfield ratio in EXPAGR or HEIA, or the ratio between management requirement and management capacity when a farming system shifts from HEIA into NCA. The following cases are illustrative of previously sustainable systems. They give clues for future development and for the choices in policy setting or research/extension priorities.

THE INFIELD/OUTFIELD SYSTEM

A mixed system that uses animals to graze "outfields" distant from the village in order to concentrate all or part of the excreted plant nutrients on the "infields" near the farmhouse is called an outfield/infield system. It is a form of mixing in the EXPAGR mode where land is abundant. The system exists (and has existed) in many forms. It can employ animals that are only stocked on crop fields at night while being grazed during the daytime, it can be intensified by constructing special stables in which straw from the crop land, litter from the forest or even topsoil from the outfield are used to conserve the nutrients of the dung and urine from the animals. As recycling on the farm increases, the system slowly moves into the LEIA or NCA mode. The system has existed throughout the world for many centuries, in pre-medieval Europe and in Russia until the early twentieth century, and it is still found today in many countries in the tropics. Improvements include the better conservation of nutrients in the stable, the better application of dung on the field (more timely and better localized), or the introduction of legumes for nitrogen supply. Examples have been given in Chapter 6, sections *Nutrient losses and their prevention* and *Leys and catch crops*, and in Photos 12, 58 and 90.

THE KANO CLOSE-SETTLED ZONE [6]

The Kano close-settled zone in the Soedano-Sahelian region of Nigeria has been the site of an intensive mixed farming system of the NCA mode where for the last 30 years all available land has been under annual cultivation. Small ruminants consume crop residues, in particular those of groundnuts and cowpeas, which are good quality fodder. The nitrogen in the residues of leguminous crops is conserved in the manure, which is transported with compound waste back to farmers' fields for use as fertilizer. Legume grains are sold, earning cash, which farmers may use to purchase inorganic fertilizer if they wish or other goods. Nutrients are added to the system when harmattan dust (a wind from the north carrying sand from the Sahara and North-Sahel) is deposited on farmers' fields during the dry season. The system allows farmers to manage an efficient nutrient recycling system centred on small ruminants and based on high labour inputs by farmers. They must keep animals tethered within the compound during the rainy season, collect crop residues and weeds for fodder, and transport the manure back to the fields.

Several elements have been important in the development of this farming system:

- The high population density provides labour and a source of agricultural innovation combined with a remarkably high density of small ruminants and donkeys.
- Land tenure is organized such that farmers have usufruct rights over the land they farm.
- The settlements are dispersed and farmers live relatively close to their fields, so as to minimize the time needed to reach the fields and to transport crop residues and manure between fields and the compound.
- The farmers use the organic fertilizer on all their fields.
- The investment in oxen, ploughs and cultivators by a few farmers has been viable because they can hire out the ploughing and cultivation service to other farmers.

In addition, the farmers engage in crop, livestock and tree production to diversify their income. Many also practise non-farming economic activities, especially

[6] Based on Harris (1996).

during the long dry season. This diversity allows them to cope with risks, whether environmental (e.g. drought) or economic (e.g. price fluctuations). The dependency on external inputs is low, permitting farmers to be independent of the fluctuations in the economy. Farmers use little inorganic fertilizer, and they maintain seed lines of favoured indigenous cultivars, rather than using the seed of commercially supplied high-yielding varieties. Hence, despite being one of the most densely populated areas in semi-arid West Africa, the system is both productive and sustainable – an example of farming in the NCA mode. The increasing population densities and labour availability are for the time being essential for a process of agricultural intensification, such as the increasing use of crop residues to feed livestock, and the use of farmyard manure on the fields.

THE MACHAKOS CASE IN KENYA[7]

A widely quoted example of population pressure and its impact on development and environment is that of the Machakos district in Kenya, another case of a shift to NCA. The key to the success of this system lies in the variety of livestock feeding methods. In the past, possibilities to integrate crops and livestock were neglected and the contribution of livestock to the household income was limited. One way of improvement was to establish individual titles to land, visualized in demarcation and enclosure of grazing areas. Subsequently, some farmers developed the grazing areas to provide grazing, timber and fuel. They used multipurpose animals, they did not aim at fast maturity, they even accepted seasonal weight losses of their livestock (a typical approach within LEIA and NCA), but they aimed at high production on an area basis (an illustration of the concept of communal ideotype). High stocking rates could be maintained through the use of crop residues, i.e. increasing population led to a reduced area of grazing land, a change in the role of cattle, and the replacement of livestock by specialized crops as the main source of cash. Adjustment was the key to sustainability, and population growth itself spontaneously responded to changed economic conditions between 1979 and 1989. Extreme shortages of land in parts of the district, combined with a national economic recession, high costs of education and other expenses for raising children, led to voluntary family planning. Finally, a programme in which people that had migrated to the city sent money back to the villages supported the process.

The evolution of the Machakos system differed between the drier and wetter areas of the district, and it depended on farm size:

- In the semi-arid region, animal draught power was important in various water conservation techniques, reducing risks of crop failure. Farmers on smaller farms combined wet season grazing with dry season use of crop residues, while larger farms relied more on grazing.
- In the subhumid region, where high population density leads to very small farms, draught power was no longer necessary and cattle were valued more for manure and milk.

In general, as land is scarce, fodder production was combined with soil conservation and stall feeding or tethering (Photo 102). Other methods of range improvement such as hedging, fencing, bush and indigenous tree management and scratch ploughing (superficial ploughing of the land), became attractive because they needed labour but almost no cash (a case of farming in LEIA mode).

Increased population led, via higher food demand, more labour, increased knowledge, and a reduction in the per capita cost of physical and social infrastructure, to an autonomous development towards higher agricultural production. The means and incentives to invest money and work in farm improvement and the knowledge of new and appropriate technologies were instrumental in this change. The new technologies came from various sources, such as traders, merchants, research and extension, religious groups, educated relatives and experimenting fellow-farmers. Literacy and general knowledge were increasingly useful to find non-agricultural work, to make the most of a farming

PHOTO 102
A bird's-eye view of land conservation in Machakos (Kenya)

---

[7] Based on Slingerland (2000), Tiffen, Mortimer and Ackello-Ogutu (1993) and Tiffen, Mortimer and Gichuki (1994a and b).

enterprise, and to participate in the various social and commercial networks. The development was assisted by government interventions and policies with respect to pricing, investments and education. Government also provided community development services and changed research and extension from a top-down approach to one involving farmers in the development of technology. It supported small towns in creating jobs and infrastructure and becoming centres of trade and services for the rural area. The Machakos case illustrates that farmers must be offered a variety of techniques from which they can select for managing pasture and other systems of animal feeding. The differentiation is necessary not only because agroclimatic conditions vary, but also because different combinations of grazing and stall-feeding may be economic for both small and large farmers.

THE FLEMISH/NORFOLK SYSTEM

This very intensive mode of integrated crop livestock farming originated in Flanders (Belgium) during the thirteenth and fourteenth centuries; it was elaborated in the United Kingdom in the seventeenth century. The system originates from conditions where the outfield/infield ratio declined due to increased population pressure. Land became scarce relative to the population and EXPAGR was replaced by NCA. Grazing of animals on the outfield to manure the infield became impossible and farmers were forced to find other ways of providing nutrients to their crops. They did so by keeping animals in a system of zero grazing on deep litter, quite similar to

PHOTO 103
The use of a mustard/berseem (cruciferae/legume) mixture for animal feed in the Gangetic plains of India to maintain nitrogen, phosphorus and soil organic matter

the Kano case. The animals were fed on crop residues and on crops that could fix or mobilize nutrients such as nitrogen and phosphorus. Legumes fixed atmospheric nitrogen and rape crops (cruciferae) were used to mobilize soil phosphorus (Figure 2 and Photos 87 and 90). The proximity of the city helped to generate cash from the sale of animal produce. Moreover, the use of night soil from the city on the crop fields was one additional way to maintain soil fertility by further recycling, i.e. integration and mixed farming at regional level.

THE MEDITERRANEAN LEGUME-GRAIN ROTATION

The effective use of legumes for nitrogen fixation in tropical areas is limited. The organic matter and associated protein quickly decomposes in conditions of high humidity and temperature, resulting in leaching and volatilization of nitrogen compounds before the crop can use this valuable nutrient. Eventually this can even lead to acidification of the soils, an example of apparently sustainable systems that eventually outdo themselves. However, the Mediterranean climate has a peculiar characteristic that distinguishes it from hot and humid tropical ones. The Mediterranean climate is characterized by wet, cool winters and hot but dry summers; it occurs in the United States and southern Australia, as well as around the Mediterranean. Under these conditions there is hardly any release of nitrogen in summer (soil microbes are inactive due to lack of water), and slow release of nitrogen in winter (water is available but temperatures are relative low). In southern Australia this resulted in a sustainable crop rotation of grains and legumes, where animals feed on the biomass of legumes, straw and failed crops.

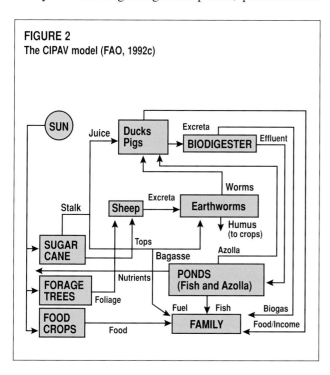

FIGURE 2
The CIPAV model (FAO, 1992c)

THE CIPAV SYSTEM[8]

A specific crop-livestock system based on a contribution of sugar cane/pigs and biogas has been developed in Colombia. It is called the CIPAV system, after the name of the institute where it was developed. It is being applied and tested in several parts of the world and the main ingredient is sugar cane, which produces the feed (juice and tops) and fuel (bagasse). Multipurpose trees and water plants such as duckweed supply the protein, while the trees also play other important roles such as controlling erosion, providing sinks for carbon dioxide and methane and as a source of biodiversity. Sugar cane and trees have well-developed systems of biological pest control; they can also do without much synthetic chemical input. Also they are easily separated into high and low-fibre fractions as required for the different end uses of feed for pigs, cattle and sheep and fuel. The preferred animal species are pigs and ducks, which adapt readily to the high-moisture feed resources (cane juice, tree leaves and water plants), and they have a high-meat:methane production ratio. Sheep, which can derive most of their feed from the cane tops and tree foliage, complement them. Buffaloes and/or triple purpose cattle can supply draught power as well as meat and milk. All the livestock are managed in partial or total confinement to minimize environmental damage

---

[8] Based on FAO (1992c).

and to maximize nutrient recycling to the crops. The CIPAV model is flexible, as witnessed by the increasing acceptance of many of the elements by resource-poor and entrepreneurial farmers.

CONCLUDING REMARKS

This chapter has discussed cases of mixed crop-livestock systems that have proven to be sustainable for shorter or longer periods of time, across the world. More systems could have been elaborated, e.g. the integrated fish/pig/vegetable systems from China (BOSTID, 1981); cattle grazing under coconuts; the multilayer system developed on Bali (Nitis, 1995); the alley farming systems (Kang, Wilson and Lawson, 1984); the traditional paddy/livestock systems of Southeast Asia, etc. The point is, however, that a variety of mixed systems can be instrumental to sustainability. Still they may have to change their mode of farming with the ratio between population pressure and resource supply, which requires change in the community organization as well as attention to the whole rather than maximum yield of individual crop components. Sustainability therefore combines an element of change and attention to relationships between systems. Particularly the latter aspect tends to be forgotten in specialized farming of the HEIA mode which, therefore, may now have to re-integrate lessons and methods that it can learn from mixed farming.

Chapter 9

Concluding comments

Mixed farming systems occur on different levels and at different intensities and in many different forms. They can be defined according to different components, i.e. on-farm, between farms, between crops, between animals and crops and, as in this document, in particular as crop-livestock systems. Even the crop-livestock systems occur in different modes, depending on the access of farmers to land, labour and capital. Individual farmers will feel no incentive towards mixed crop-livestock farming if either cropping or raising livestock is impossible due to factors such as unreliable rainfall, low soil fertility or presence of particular pests and diseases. Farmers will also not be inclined to apply mixed farming when conditions favour specialization but they can become interested in mixed crop-livestock farming when mixing helps to overcome local problems of soil fertility, resource-depletion or waste-disposal. Mixed farming is not just a thing of the past; modern high-input systems can also learn a lot from the older mixed systems, both in terms of methods of integrating crops and livestock, as well as in terms of obtaining new ways of looking at agricultural productivity. The issue of the communal ideotype is particularly important where components of the farm need to be mutually adjusted. A high yield of the total system is to set the limit of the yield of the components. In that same sense, a lower yield of components may give a higher yield of the total. Highlighting these issues may not only serve to achieve a new approach to farming but may also set an example for other sectors of society.

References and suggested reading

Aarts, H.F.M., Habekotté, B. & van Keulen, H. 2000. Nitrogen (N) management in the 'De Marke' dairy farming system. *Nutr. Cycl. Agroecosyst.*, 56: 231-240.

Abrahamson, W.G. ed. 1989. *Plant-animal interactions.* New York, McGraw-Hill. 481 pp.

ADB. 1991. *Sector papers on livestock.* Staff Paper 4. Manila, Asian Development Bank (ADB).

Altieri, M. 1991, How best can we use biodiversity in agro-ecosystems? *Outlook on Agriculture*, 20(1): 15-23.

Andreson, N. 2000. *The foraging pig. Resource utilisation, interaction, performance and behaviour of pigs in cropping systems.* Agraria 227. Swedish University of Agricultural Sciences, Uppsala, Sweden. (Ph.D. thesis)

Bayer, W., Sulieman, R., Kaufmann, R.V. & Waters-Bayer, A. 1987. Resource use and strategies for development of pastoral systems in subhumid West Africa – the case of Nigeria. *Quarterly J. Int. Agric.*, 26: 58-71.

Blood, D.C., Henderson, J.A. & Radostis, O.M. 1979. *Veterinary medicine.* Fifth edition. London, The English Language Book Society and Bailliere Tindall. 1135 pp.

Bosman, H.G. 1995, *Productivity assessments in small ruminant improvement programmes. A case study of the West African dwarf goat.* Wageningen Agricultural University, Wageningen, the Netherlands. 224 pp. (Ph.D. thesis)

BOSTID. 1981. *Food, fuel and fertilizer from organic wastes.* Report of an ad hoc panel of the Advisory Committee on Technology Innovation. Board on Science and Technology for International Development (BOSTID). Washington, DC, National Academy Press. 154 pp.

Breman, H. & De Wit, C.T. 1986. Rangeland productivity and exploitation in the Sahel. *Science*, 221(4618): 1341-1347.

Briggs, M.H. ed. 1996. *Urea as a protein supplement.* Oxford, UK, Permagon Press. 466 pp.

Campbell, C.A. 1996. *In* A. Budelman, ed. *Agricultural R&D at the crossroads: merging systems research and social actor approaches.* Amsterdam, Royal Tropical Institute. p. 169-184

Centro Mesoamericano de Estudios sobre Tecnología Apropiada (CEMAT). 1987. Memorias. Primer seminario-taller nacional sobre letrinas aboneras secas familiares. 144 pp.

Chen, H. *In* H. Hayakawa, M. Sasaki & K. Kimura, eds. *Integrated systems of animal production in the Asian region.* Proceedings of a symposium held in conjunction with the 8th AAAP Animal Science Congress, Chiba, Japan, 13-18 October 1996. AAAP and FAO, Rome.

CIRAN/NUFFIC. *Indigenous knowledge and development monitor.* Published three times a year. Magazine produced by the Centre for International Research and Advisory Networks (CIRAN/NUFFIC), the Netherlands.

Comaroff, J. 1992, Goodly beasts and beastly goods. *In* J. Comaroff & J. Comaroff. *Ethnography and the historical imagination*, Chapter 5. Boulder, Westview Press, Boulder. 333 pp.

Conway, G.R. & Barbier, E.B. 1990. *After the green revolution. Sustainable agriculture for development*, p. 131-137. London, Earthscan Publications.

Coppock, D.L., Ellis, J.E. & Swift, D.M. 1986. Livestock feeding ecology and resource utilization in a nomadic pastoral ecosystem. *J. Appl. Ecol.*, 23: 573-583.

Crotty, R. 1980. *Cattle, economics and development.* UK, Commonwealth Agricultural Bureaux. 253 pp.

Defoer, T., Budelman, A., Toulmin, C. & Carter, S.E. eds. 2000. Building common knowledge: participatory learning and action research (part 1). *In* T. Defoer & A. Budelman, eds. *Managing soil fertility in the tropics. A resource guide for participatory learning and action research.* Royal Tropical Institute, Amsterdam.

De Haan, C., Steinfeld, H. & Blackburn, H. 1997. *Livestock and the environment: finding a balance.* Fressingfield, Eye, Suffolk, UK, WREN Media. 115 pp.

De Jong, R. 1996. *Dairy stock development and milk production with smallholders.* Wageningen Agricultural University, the Netherlands. 308 pp. (Ph.D. thesis)

Dembele, I., Kone, D., Soumare, A., Coulibaly, D., Kone, Y., Ly, B. & Kater, L. 2000. Fallows and field systems in dryland Mali. *In* T. Hilhorst & F. Muchena, eds. *Nutrients on the move. Soil fertility dynamics in African farming systems*, p. 83-101. London, IIED.

Devendra, C. & Burns, M. 1970. *Goat production in the tropics.* Technical Communication No. 19. Commonwealth Bureau of Animal Breeding and Genetics. Commonwealth Agricultural Bureaux, Farnham Royal, UK. 184 pp.

Devendra, C. & Sevilla, C. eds. 1995. *Crop-animal interaction*. Proceedings of an international workshop. International Rice Research Institute, Manila. 574 pp.

Domenech, J. 1994. *Importance des mandats sanitaires vétérinaires dans le processus de privatisation de la profession. Résultats obtenus au Tchad en 1994*. Présenté à l'Atelier internationale de Bamako sur le rôle des entreprises dans le processus de la privatisation des services vétérinaires en Afrique francophone. 5-8 décembre 1994. Bamako.

Donald, C.M. 1981. Competitive plants, communal plants, and yield in wheat crops. *In* L.T. Evans & W.J. Peacock, eds. *Wheat science – today and tomorrow*, p. 243-247. Cambridge, UK, Cambridge University Press. 290 pp.

Doyle, P.T., Pearce, G.R. & Egan, A.R. 1986. Potential of cereal straws in tropical and temperate regions. p. 63-79. *In* M.N.M. Ibrahim & J.B. Schiere, eds. *Rice straw and related feeds in ruminant rations*. Proceedings of an international workshop held in Kandy, Sri Lanka, 24-28 March 1986. Straw Utilization Project, Kandy, Sri Lanka.

EC. 1997. *Evaluation of the Pan-African Rinderpest Campaign*. Final Report. Brussels, European Commission, DGVIII.

Efde, S.L. 1996. *Quantified and integrated crop and livestock production analysis at farm level*. Wageningen, the Netherlands, Wageningen Agricultural University. (Ph.D. thesis)

Euroconsult. 1989. *Agricultural compendium for rural development in the tropics and the subtropics*. Amsterdam, Elsevier. 740 pp.

FAO. 1978. Utilizing breed differences in growth of cattle in the tropics, by J.E. Frisch & J.E. Vercoe. In *World Animal Review*, 25: 8-12.

FAO. 1983. *Integrating crops and livestock in West Africa*. FAO Animal Production and Health Paper No. 41. Rome. 112 pp.

FAO. 1988. Feeding of urea/-ammonia treated rice straw, by J.B. Schiere, A.J. Nell M.N.M. Ibrahim. In *World Animal Review*, 65: 31-42.

FAO. 1992a. Feeding systems based on traditional use of trees for feeding livestock, by D.V. Rangnekar. *In* A. Speedy & P. Pugliese, eds. *Legume trees and other fodder trees as protein sources for livestock*. FAO Animal Production and Health Paper No. 102. Rome. 339 pp.

FAO. 1992b. The role of fodder trees in Philippine smallholder farms, by F.A. Moog. *In* A. Speedy & P. Pugliese, eds. *Legume trees and other fodder trees as protein sources for livestock*. FAO Animal Production and Health Paper No. 102. Rome. 339 pp.

FAO. 1992c. The role of multipurpose trees in integrated farming systems for the wet tropics, by T.R. Preston. *In* A. Speedy & P. Pugliese, eds. *Legume trees and other fodder trees as protein sources for livestock*. FAO Animal Production and Health Paper No. 102. Rome. 339 pp.

FAO. 1992d. *Legume trees and other fodder trees as protein sources for livestock*. A. Speedy & P. Pugliese, eds. FAO Animal Production and Health Paper No. 102. Rome, 339 pp.

FAO. 1995a. Livestock, a driving force for food security and sustainable development, by R. Sansoucy. *World Animal Review*, 84/85: 5-17.

FAO. 1995b. FAO strategy for international animal health. p. 25-31. *World Animal Review*, 84/85: 25-31.

FAO. 1995c. Emergency prevention system for transboundary animal and plant diseases: the livestock diseases component, by M.M. Rweyemamu, P.L. Roeder, A. Benkirane, K. Wojciechowski & A. Kamata. *World Animal Review*, 84/85: 74-82.

FAO. 1995d. *World livestock production systems. Current status, issues and trends*, by C. Sere & H. Steinfeld. Rome.

FAO. 1995e. Livestock's contribution to the protection of the environment, by C. Dalibard. *World Animal Review*, 84/85: 104-112.

FAO. 1995f. Analysis of current trends in the distribution patterns of ruminant livestock in tropical Africa, by P.N. De Leeuw & B. Rey. *World Animal Review*, 83: 47-50.

FAO. 1996. Multipurpose use of animals. Comment, by J.C. Chirgwin. *World Animal Review*, 86.

FAO. 1998. *Area-wide integration of crop-livestock activities*, by Y.W. Ho & Y.K. Chan, eds. Proceedings of a regional workshop. FAO Regional Office for Asia and the Pacific, Bangkok.

FAO. 1999. Making better use of animal resources in a rapidly urbanizing world: a professional challenge, by M. Ghirotti. *World Animal Review*, 92: 1-14.

FAO. 2000. *Peri-urban livestock systems. Problems, approaches and opportunities*, by J.B. Schiere. Report prepared for FAO Animal Production and Health Division, Rome.

Gbego, I.T. 1992. *Synthèse des recherches sur l'alimentation des caprins au plateau Adja*. Recherche appliquée en milieu réel, Lokossa, Benin. 16 pp.

Gbego, I.T. & Van den Broek, A. 1992. *La productivité des petits ruminants sur le plateau Adja et dans la depression de Tchi*. Recherche appliquée en milieu réel, Lokossa, Benin. 27 pp.

Giampietro, M. 1997. Socioeconomic constraints to farming with biodiversity. *Agriculture, Ecosystems & Environment*, 62: 145-167.

Gibson, T. 1987. A ley farming system using dairy cattle in the infertile uplands. *Northeast Thailand World Animal Review*, 61: 36-43.

Hayakawa, H., Sasaki, M. & Kimura, K. eds. 1996. *Integrated systems of animal production in the Asian region.* Proceedings of a symposium held in conjunction with the 8th AAAP Animal Science Congress, Chiba, Japan, 13-18 October 1996. AAAP and FAO, Rome. 111 pp.

Hammond, K. 2000. *A global strategy for the development of animal breeding programmes in lower-input production environments.* World Expo 2000 Workshop on Animal Breeding and Animal Genetic Resources, Institute for Animal Science and Animal Behaviour, Mariensee, Germany, 17-18 July 2000.

Harris, M. 1965. The myth of the sacred cow. *In* A. Leeds & A.P. Vayda, eds. *Man, culture and animals: the role of animals in human ecological adjustments*, p. 217-228. Based on a symposium of the American Association for the Advancement of Science, 30 December 1961, Denver, USA. Also published in a more detailed version under the title The cultural ecology of India's sacred cattle in *Current Anthropology*, 7.

Harris, F. 1996. *Intensification of agriculture in semi-arid areas: lessons from the Kano close settled zone, Nigeria.* Gatekeeper Series, No. 59. London, IIED. 20 pp.

Henzell, E.F. 1977. Nitrogen nutrition of tropical pastures. *In* P.K. Skerman, ed. *Tropical forage legumes.* FAO Plant Production and Protection Series No. 2. Rome. 609 pp.

Hilhorst, T. & Coulibaly, A. 1998. *Une convention locale pour la gestion participative de la brousse au Mali.* Dossier no. 78, Programme zones arides. London, IIED.

Hilhorst, T. & Muchena, F. eds. 2000. *Nutrients on the move. Soil fertility dynamics in African farming systems.* London, International Institute for Environment and Development. 146 pp.

Hoffland, E., Findenegg, G. & Nelemans, J. 1989. Solubilization of rock phosphate by rape I/II. *Plant and Soil*, 113: 155-165.

Holden, S. 1997. *Economic impact of community-based animal health workers in Kathakani, Kenya.* Crewkerne: Livestock in Development.

Holling, C.S. 1987a. Simplifying the complex: the paradigms of ecological function and structure. *European J. Operational Research*, 30: 139-146.

Holling, C.S. 1987b. What barriers, what bridges? *In* L.H. Gunderson, C.S. Holling & S.S. Light, eds. *Barriers and bridges to the renewal of ecosystems and institutions.* p. 3-34. New York, Columbia Press.

Holling, C.S., Berkes, F. & Folke, C. 1997. Science, sustainability and resource management. *In* F. Berkes, C.

Folke & J. Colding. eds. *Linking social and ecological systems. Management practices and social mechanisms for building resilience*, p. 342-362. Cambridge/New York/Melbourne, Cambridge University Press.

Hong, S. 1988. Traditional medical systems in East Asia. *Bibliographies in Technology and Social Change No. 3.* Technology and Social Change Program, Iowa State University, Ames, Iowa, USA.

Ifar, S. 1996. *Relevance of ruminants in upland mixed farming systems in East Java, Indonesia.* Wageningen Agricultural University, the Netherlands. 139 pp. (Ph.D. thesis)

IIRR. 1994. *Ethnoveterinary medicine in Asia: An information kit on traditional animal health care practices.* Four volumes. International Institute of Rural Reconstruction, Silang, Cavite, the Philippines.

ILRI. 1995. *Livestock policy analysis.* ILRI Training Manual 2. Addis Ababa.

Jahnke, H.E. 1982. *Livestock production systems and livestock development in tropical Africa.* Vauk, Kieler Wissenschaftsverlag. 253 pp.

Jansen, J.C.M. & de Wit, J. 1996. *Livestock and the environment. Finding a balance. Environmental impact assessment of livestock production in mixed irrigated systems in the sub-humid zones.* Wageningen, the Netherlands, IAC. 72 pp.

Joldersma, R., Coulibaly, L., Diarra, A., Hilhorst, T. & Vlaar, J. 1994. *SIWAA, La brousse sèche. Expérience de gestion villageoise d'un terroir intervillageois au Mali.* Sikasso, Mali, ESPGRN. 64 pp.

Joshi, A.L., Doyle, P.T. & Oosting, S.J. 1994. *Variation in the quantity and quality of fibrous crop residues.* Proc. National Seminar, BAIF Development Research Foundation, Pune, Maharashtra, India, 8-9 February 1994. Pune, India. 174 pp.

Kabourakis, E. 1996. *Prototyping and dissemination of ecological olive production systems.* Wageningen Agricultural University, the Netherlands. 121 pp. (Ph.D. thesis)

Kang, B.T., Wilson, G.F. & Lawson, T.L. 1984. *Alley cropping, a stable alternative to shifting cultivation.* International Institute of Tropical Agriculture, Ibadan, Nigeria.

Kater, L. 2000. Fallows and field systems in dryland Mail. *In* T. Milhorst & F. Machena, eds. *Nutrients on the move. Soil fertility dynamics in African farming systems.* International Institute for Environment and Development, London. 146 pp.

Kater, L., Dembele, I., Kone, D., Brock, K. & Budelman, A. 2000. Millet farming in central Mali. *In* A.

Budelman & T. Defoer, eds. *PLAR and resource flow analysis in practice. Case studies from Benin, Ethiopia, Kenya, Mali and Tanzania*, p. 71-93 (Part 2). Amsterdam, the Netherlands, Royal Tropical Institute.

Kay, J.J., Regier, H.A., Boyle, M.B. & Francis, G. 1999. An ecosystem approach to sustainability: addressing the challenge of complexity. *Futures*, 31: 721-742.

Kidane, H. 1984. *The economics of farming systems in different ecological zones of the Embu district (Kenya), with special reference to dairy production*. Hannover University, Hannover. 227 pp. (Ph.D. thesis)

Lans, C. & Brown, G. 1998. Ethnoveterinary medicines used for ruminants in Trinidad and Tobago. *Prev. Vet. Med.*, 35(3): 149-163.

Loomis, R.S. & Connor, D.J. 1992. *Crop ecology: productivity and management in agricultural systems*. Cambridge, UK, Cambridge University Press. 538 pp.

Lord Ernle. 1961. *English farming, past and present*. Sixth edition. London/Melbourne/Toronto, Heinemann and London, Frank Cass. 559 pp.

Mäki-Hokkonen, J. 1996. Integrated systems of animal production in Asian region FAO's studies into Asian livestock production systems; FAO's programme priorities. *In* H. Hayakawa, M. Sasaki & K. Kimura, eds. *Integrated systems of animal production in the Asian region*, p. 1-8. Proceedings of a symposium held in conjunction with the 8th AAAP Animal Science Congress, Chiba, Japan, 13-18 October 1996. AAAP and FAO, Rome.

Morrison, D.A., Kingwell, R.S., Pannell, D.J. & Ewing, M.A. 1986. A mathematical programming model of a crop-livestock farm system. *Agric. Syst.*, 20: 243-268.

Müller-Sämann, K.M. & Cotschi, J. 1994. *Sustaining growth. Soil fertility management in tropical smallholdings*. Weikersheim, Germany, Margraf Verlag, CTA and GTZ. 486 pp.

Mureithi, J.G., Tayler, R.S. & Thorpe, W. 1995. Productivity of alley farming with *Leucaena (Leuceana leucocephala* La. de Wit and Napier grass (*Pennisetum purpureum* K. Schum) in coastal lowland, Kenya. *Agroforestry Systems*, 31(28): 1-20.

Nitis, I.M. 1995. Research methodology for semi-arid crop-animal systems in Indonesia. *In* C. Devendra & C. Sevilla, eds. *Crop-animal interaction*, p. 301-333. IRRI Discussion Paper Series No. 6. International Rice Research Institute, Manila, the Philippines. 572 pp.

Norgaard, R.B. 1984. Coevolutionary development potentials. *Land Economics*, 60(2): 160-173.

NRC. 1989. *Alternative agriculture*. National Academic Press, Washington. Committee on the Role of Alternative Farming Methods in Modern Production Agriculture. National Research Council.

Nye, P.H. & Greenland, D.J. 1961. *The soil under shifting cultivation*. Reading, UK, Commonwealth Agricultural Bureaux.

Odum, H.T. 1983. *Systems ecology, an introduction*. New York, Wiley Interscience. 644 pp.

Oomen, G.J.M., Lantinga, E.A., Goewie, E.A. & Van Der Hoek, K.W. 1998. Mixed farming systems as a way towards a more efficient use of nitrogen in European Union agriculture. *Environmental Pollution*, 102(S1): 6697-6704.

Patil, B.R., Rangnekar, D.V. & Schiere, J.B. 1993. Modelling of crop-livestock integration: effect of choice of animals on cropping patterns. *In* K. Singh and J.B. Schiere, eds. Feeding of ruminants on fibrous crop residues: aspects of treatment, feeding, nutrient evaluation, research and extension, p. 336-343. Proc. Int. Workshop, 4-8 February 1991, NDRI-Karnal. New Delhi, ICAR.

Pearson, C.J. ed. 1992. *Ecosystems of the world – field crop ecosystems*. Amsterdam, London, New York, Tokyo, Elsevier. 570 pp.

Pingali, P., Bigot, Y. & Binswanger, H.P. 1987. *Agricultural mechanization and the evolution of systems in sub-Saharan Africa*. World Bank, The John Hopkins University Press, Baltimore and London. 216 pp.

Powell, M. 1986. Manure for cropping: a case study from central Nigeria. *Expl. Agriculture*, 22: 15-24.

Powell, J.M. & Waters-Bayer, A. 1995. Interactions between livestock husbandry and cropping in a West African Savanna. *In* J.C. Tothill & J.J. Mott, eds. *Ecology and management of the world's savannas*, p. 252-255. Canberra, The Australian Academy of Sciences.

Powell, J.M. & Williams, T.O. 1993. *Livestock, nutrient cycling and sustainable agriculture in West Africa*. Gatekeeper series No. 37. London, IIED. 15 pp.

Powell, J.M., Fernández-Rivera, S., Williams, T.O. & Renard, C. eds. 1993. *Livestock and sustainable nutrient cycling in mixed farming systems of sub-Saharan Africa*. Proc. Intern. Conf., 22-26 November1993. International Livestock Centre for Africa (ILCA), Addis Ababa.

Preston, T.R. & Leng, R.A. 1984. Supplementation of diets based on fibrous residues and by-products. *In* R.F. Sundstol & E. Owen, eds. *Straw and other fibrous by-products as feed*, p. 373-413. Development in Animal and Veterinary Sciences, 14. Amsterdam, Elsevier.

Pretty, J.N. 1995. *Regenerating agriculture. Policies and practice for sustainability and self-reliance*. London, Earthscan. 320 pp.

Rangnekar, D.V., Badve, V.C., Kharat, S.T., Sobale, B.N. & Joshi A.L. 1982. Effect of high pressure steam treatment on chemical composition and digestibility *in vitro* of roughages. *Animal Feed Science and Technology,* 7: 61-70.

Reijntjes, C., Haverkort, B. & Waters-Bayer, A. 1992. *Farming for the future. An introduction to low-external input and sustainable agriculture.* Leusden, the Netherlands, Macmillan, ILEIA. 250 pp.

Renard, C. ed. 1997. *Crop residues in sustainable mixed crop/livestock farming systems.* New York, CAB International. 322 pp.

Reeves, T.G. & Ewing, M.A. 1993. Is ley farming in mediterranean zones just a passing phase? *In* Proceedings of the International Grassland Congress. SIR Publishing, Wellington, New Zealand.

Roberts, K. & Coutts, J. 1997. *A broader approach to common resource management: Landcare and Integrated Catchment Management in Queensland, Australia.* Agricultural Research and Extension Network Paper No. 70. London, Overseas Development Institute.

Robineau, L. ed. 1991. *Towards a Caribbean ethno-pharmacopoeia. Scientific research and popular use of medicinal plants in the Caribbean.* Santo Domingo, DO, Enda-caribe, UNAH.

Röling, N. 1996. Towards an interactive agricultural science. *European Journal of Agricultural Education and Extension,* 2: 35-48.

Sato, K., Sansoucy, R. & Preston, T.R. 1996. FAO regional project on "better use of locally available feed resources in sustainable agriculture system". *In* H. Hayakawa, M. Sasaki, & K. Kimura, eds. *Integrated systems of animal production in the Asian region,* p. 23-30. Proceedings of a symposium held in conjunction with the 8th AAAP Animal Science Congress, Chiba, Japan, 13-18 October 1996. AAAP and FAO, Rome.

Savadogo, M. 2000. *Crop residue management in relation to sustainable land use. A case study in Burkina Faso.* Wageningen University, Wageningen, the Netherlands. (Ph.D. thesis)

Schiere, J.B. 1995. *Cattle, straw and system control.* Wageningen Agricultural University, Royal Tropical Institute, Amsterdam. 216 pp. (Ph.D. thesis)

Schiere, J.B. & De Wit, J. 1993. Feeding standards and feeding systems. *Animal Feed Science & Technology,* 43: 121-134.

Schiere, J.B. & De Wit, J. 1995. Feeding of urea ammonia treated straw in the tropics. Part II: Assumption on nutritive values and their validity for least cost ration formulation. *Animal Feed Science & Technology,* 51: 45-63.

Schiere, J.B. & van Keulen, H. 1999. Rethinking high input systems of livestock production: a case study of nitrogen emissions in Dutch dairy farming. *Tropical Grasslands,* 33: 1-10.

Schiere, J.B., Ibrahim, M.N.M. & van Keulen, H. 2001. Sustainability through mixed crop-livestock farming. A review of definitions, criteria and scenario studies under varying resource fluxes. *J. Agriculture, Ecosystems and Environment* (in press).

Schiere, J.B., Kiran Singh, & De Boer, A.J. 2000. Farming systems research applied in a project on feeding of crop residues in India. *Expl. Agric.,* 36: 51-62.

Schiere, J.B., Steenstra, F., De Wit, J. & van Keulen, H. 1999. The effect of restricted input use on the design of farming systems; scenario studies about feed allocation for maximum livestock system output. *Netherlands J. Agric. Sci.,* 47: 169-183.

Schoonmaker Freudenberger, K., Wood, E., Dieye, A.M., Diop, P.M., Drame, M., Ka, F., Ly, M., N'diaye, D., Tappan, G. & Thiam, M. 2000. Cattle and cultivators: conflicting strategies for natural resource management. *LEISA,* 16(1): 8-10.

Scoones, I., ed. 1996. *Living with uncertainty: new directions in pastoral development in Africa.* London, Intermediate Technology Publications. 210 pp.

Seetharam, A., Subba Rao, A. & Schiere, J.B. eds. 1995. *Crop improvement and its impact of the feeding value of straw and stovers of grain cereals in India.* Proceedings of a workshop held at Krishi Bhavan, 21 November 1994. Indian Council of Agricultural Research, New Delhi.

Singh, K. & Schiere, J.B. eds. 1993. *Feeding of ruminants on fibrous crop residues. Aspects of treatment, feeding, nutrient evaluation, research and extension.* Indian Council of Agricultual Research (Animal Sciences) (ICAR), Krishi Bhavan, New Delhi and Department of Tropical Animal Husbandry, Wageningen Agricultural University, the Netherlands. 486 pp.

Singh, K. & Schiere, J.B. eds. 1995. *Handbook for straw feeding systems. Principles and applications with emphasis on Indian livestock production.* ICAR, Krishi Bhavan, New Delhi, and Department of Animal Production Systems, Wageningen Agricultural University, the Netherlands. 428 pp.

Slingerland, M. 2000. *Mixed farming: scope and constraints in West African savanna.* Wageningen, the Netherlands, Wageningen Institute of Animal Sciences. 289 pp. (thesis)

Spedding, C.R.W. 1979. *An introduction to agricultural systems.* Second edition, Amsterdam, Elsevier Applied Science. 225 pp.

Starkey, P. 1996. *Networking for sustainable agriculture: lessons from animal traction development.* Gatekeeper Series No. 58. London, IIED. 18 pp.

Starkey, P., Mwenya, E. & Stares, J. 1992. *Improving animal traction technology.* Proceedings of the Animal Traction Network for Eastern and Southern Africa (ATNESA) workshop held 18-23 January 1992, Lusaka, Zambia. Technical Centre for Agricultural and Rural Cooperation (CTA), Wageningen, the Netherlands.

Sumberg, J. 1996. *Food production in and around urban areas: A bibliography with particular reference to livestock, Tanzania and sub-Saharan Africa.* School of Development Studies, University of East Anglia, Norwich, UK.

Sumberg, J. 1997. Policy, milk and the Dar es Salaam peri-urban zone: a new future for an old development theme? *Land Use Policy* 14(4): 277-293.

Sumberg, J. 1998a. The Dar es Salaam milk system: dynamics of change and sustainability. *Habitat International*, 23(2): 189-200.

Sumberg, J. 1998b. Poultry production in and around Dar es Salaam, Tanzania: competition and complementarity. *Outlook on Agriculture*, 27(3): 177-185.

Sundstøl, F. & Owen, E. eds. 1984. *Straw and other fibrous by-products as feed.* Developments in Animal and Veterinary Sciences 14. Elsevier, Amsterdam. 604 pp.

Tanner, J.C., Holden, S.J., Winugroho, M., Owen, E. & Gill, M. 1995. Feeding livestock for compost production: A strategy for sustainable upland agriculture on Java. *In* J.M. Powell, S. Fernández-Rivera, T.O. Williams & C. Renard, eds. *Livestock and sustainable nutrient cycling in mixed farming systems of sub-Saharan Africa.* Proc. Intern. Conf., 22-26 Nov. 1993, International Livestock Centre for Africa (ILCA), Addis Ababa.

Theron, J.J. & Haylett, D.G. 1953. The regeneration of soil humus under a grass ley. *Emp. J. Exp. Agric.*, 21: 81-93.

Thomas, R.J. & Lascano, C.E. 1995. The benefits of forage legumes for livestock production and nutrient cycling in pasture and agropastoral systems of acid-soil savannahs of Latin-America. *In* J.M. Powell, S. Fernández-Rivera, T.O. Williams & C. Renard, eds. *Livestock and sustainable nutrient cycling in mixed farming systems of sub-Saharan Africa.* Proc. Intern. Conf., 22-26 November 1993, International Livestock Centre for Africa (ILCA), Addis Ababa.

Tiffen, M., Mortimer, M. & Ackello-Ogutu, A.C. 1993. *From agropastoralism to mixed farming: the evolution of farming systems in Machakos, Kenya, 1930-1990.* Agricultural Administration (Research and Extension) Network Paper 45. London, Overseas Development Institute. 35 pp.

Tiffen, M., Mortimer, M. & Gichuki, F. 1994a. *More people, less erosion, environmental recovery in Kenya.* Overseas Development Institute. Chichester, John Wiley. 311 pp.

Tiffen, M., Mortimer, M. & Gichuki, F. 1994b. *Population growth and environmental recovery, policy lessons from Kenya.* Gatekeeper Series 45. London, International Institute for Environment and Development. 26 pp.

Tulachan, P.M. & Neupane, A. 1999. *Livestock in mixed farming systems of the Hindu Kush-Himalayas. Trends and sustainability.* International Centre for Integrated Mountain Development, Kathmandu. 116 pp.

Udo, H.M.J. 1997. Relevance of farmyard animals to development. *Outlook on Agriculture*, 26(1): 25-28.

Udo, H.M.J. 1999. *Management of genetic resources.* Chapter 4 in draft lecture notes on Introduction of livestock systems, J.B. Schiere & H.N.M. Ibrahim, Wageningen Agricultural University, the Netherlands.

Upton, M. 1985. Models for improved production systems for small ruminants. In *Sheep and goats in humid West-Africa.* Proc. ILCA Workshop, Ibadan, January 1984.

Utomo, B.N. 2001. *Potential of oil palm solid wastes as local feed resource for cattle in central Kalimantan, Indonesia.* Animal Production System Group, Wageningen University, the Netherlands. (M.Sc. thesis)

Van der Lee, J., Udo, H.M.J. & Brouwer, B.O. 1993. Design and validation of an animal traction module for a smallholder livestock systems simulation model. *Agric. Systems*, 43: 199-227.

Vandermeer, J., Van Noordwijk, M., Anderson. J., Chin Ong & Perfecto, Y. 1998. Global change and multi species agroecosystems: concepts and issues. *Agriculture, Ecosystems and Environment*, 67: 1-22.

Van Der Pol, F. 1992. *Soil mining: an unseen contributor to farm income in Southern Mali.* Bull. 325. Amsterdam, Royal Tropical Institute. 48 pp.

Vereijken, P. 1997. A methodical way of prototyping integrated and ecological arable farming systems (I/EAFS) in interaction with pilot farms. *European J. Agronomy*, 7: 235-250.

Wahed, R.A., Owen, E., Naate, M. & Hosking, B.J. 1990. Feeding straw to small ruminants. Effect of amount offered on intake and selection of barley straw by goats and sheep. *Anim. Prod.*, 51: 283-289.

Walaga, C., Egulu, B., Bekunda, M. & Ebanyat, P. 2000. Impact of policy change on soil fertility management in Uganda. *In* T. Hilhorst & F. Muchena, eds. *Nutrients on* the move. *Soil fertility dynamics in African farming systems*. London, International Institute for Environment and Development.146 pp.

Walshe, M.J. 1985. Investment for sustainable livestock development in developing countries. *In* S. Mack, ed. *Strategies for sustainable animal agriculture in developing countries*. Proceedings of the FAO Expert Consultation, Rome, 10-14 December 1990.

WCED. 1987. *Our common future*. Oxford, UK/New York, Oxford University Press, for World Commission on Environment and Development. 400 pp.

Webster, C.C. & Wilson, P.N. 1980. *Agriculture in the tropics, second edition*. Tropical Agriculture Series. New York, Longman Scientific & Technical. 640 pp.

Wilkinson, R.G. 1973. *Poverty and progress: an ecological model of economic development*. London, Methuen. 225 pp.

Winrock International. 1992. *Assessment of animal agriculture in sub-Saharan Africa*. Winrock International Livestock Research and Training Centre, Morrilton, Arkansas, USA.

Zemmelink, G. 1980. *Effect of selective consumption on voluntary intake and digestibility of tropical forages*. Doctoral thesis. Agric. Res. Pub. 896. Wageningen, the Netherlands, Pudoc. 100 pp.

FAO TECHNICAL PAPERS

FAO ANIMAL PRODUCTION AND HEALTH PAPERS

138	Producción de cuyes (*Cavia porcellus*), 1997 (S)
139	Tree foliage in ruminant nutrition, 1997 (E)
140/1	Analisis de sistemas de producción animal – Tomo 1: Las bases conceptuales, 1997 (S)
140/2	Analisis de sistemas de producción animal – Tomo 2: Las herramientas basicas, 1997 (S)
141	Biological control of gastro-intestinal nematodes of ruminants using predacious fungi, 1998 (E)
142	Village chicken production systems in rural Africa – Household food security and gender issues,1998 (E)
142	Village chicken production systems in rural Africa – Household food security and gender issues,1998 (E)
143	Agroforestería para la producción animal en América Latina, 1999 (S)
144	Ostrich production systems, 1999 (E)
145	New technologies in the fight against transboundary animal diseases, 1999 (E)
146	El burro como animal de trabajo – Manual de capacitación, 2000 (S)
147	Mulberry for animal production, 2001 (E)
148	Los cerdos locales en los sistemas tradicionales de producción, 2001 (S)
149	Animal production based on crop residues – Chinese experiences, 2001 (C)
150	Pastoralism in the new millenium, 2001 (E)
151	Livestock keeping in urban areas – A review of traditional technologies based on literature and field experiences, 2001 (E)
152	Mixed crop-livestock farming – A review of traditional technologies based on literature and field experiences, 2001 (E)

Availability: December 2001

Ar	–	Arabic	Multil	–	Multilingual
C	–	Chinese	*		Out of print
E	–	English	**		In preparation
F	–	French			
P	–	Portuguese			
S	–	Spanish			

The FAO Technical Papers are available through the authorized FAO Sales Agents or directly from Sales and Marketing Group, FAO, Viale delle Terme di Caracalla, 00100 Rome, Italy.